HOW TO BUILD & POWER TUNE
HOLLEY
CARBURETORS

Also From Veloce Publishing:

SpeedPro Series
4-Cylinder Engine - How to Blueprint & Build a Short Block for High Performance by Des Hammill
Alfa Romeo Twin Cam Engines - How to Power Tune by Jim Kartalamakis
BMC 998cc A-Series Engine - How to Power Tune by Des Hammill
BMC/Rover 1275cc A-Series Engines - How to Power Tune by Des Hammill
Camshafts - How to Choose & Time them for Maximum Power by Des Hammill
Cylinder Heads - How to Build, Modify & Power Tune Updated & Revised Edition by Peter Burgess
Distributor-type Ignition Systems - How to Build & Power Tune by Des Hammill
Fast Road Car - How to Plan and Build New Edition by Daniel Stapleton
Ford SOHC 'Pinto' & Sierra Cosworth DOHC Engines - How to Power Tune Updated & Enlarged Edition by Des Hammill
Ford V8 - How to Power Tune Small Block Engines by Des Hammill
Harley-Davidson Evolution Engines - How to Build & Power Tune by Des Hammill
Holley Carburetors - How to Build & Power Tune New Edition by Des Hammill
Jaguar XK Engines - How to Power Tune New Edition by Des Hammill
MG Midget & Austin-Healey Sprite - How to Power Tune Updated Edition by Daniel Stapleton
MGB 4-Cylinder Engine - How to Power Tune by Peter Burgess
MGB - How to Give your MGB V8 Power Updated & Revised Edition by Roger Williams
MGB, MGC & MGB V8 - How to Improve by Roger Williams
Mini Engines - How to Power Tune on a Small Budget 2nd Edition by Des Hammill
Motorsport - Getting Started in by SS Collins
Nitrous Oxide Systems - How to Build & Power Tune by Trevor Langfield
Rover V8 Engines - How to Power Tune by Des Hammill
Sportscar/Kitcar Suspension & Brakes - How to Build & Modify Enlarged & Updated 2nd Edition by Des Hammill
SU Carburettors - How to Build & Modify for High Performance by Des Hammill
Suzuki 4WD for Serious Offroad Action - Modifying by John Richardson
Tiger Avon Sportscar - How to Build Your Own Updated & Revised 2nd Edition by Jim Dudley
TR2, 3 & TR4 - How to Improve by Roger Williams
TR5, 250 & TR6 - How to Improve by Roger Williams
V8 Engine - How to Build a Short Block for High Performance by Des Hammill
Volkswagen Beetle Suspension, Brakes & Chassis - How to Modify for High Performance by James Hale
Volkswagen Bus Suspension, Brakes & Chassis - How to Modify for High Performance by James Hale
Weber DCOE & Dellorto DHLA Carburetors - How to Build & Power Tune 3rd Edition by Des Hammill

Those were the days ... Series
Alpine Trials & Rallies 1910-1973 by Martin Pfunder
Austerity Motoring by Malcolm Bobbitt
Brighton National Speed Trials by Tony Gardiner
British Police Cars by Nick Walker
Crystal Palace by Sam Collins
Dune Buggy Phenomenon by James Hale
More Dune Buggies by James Hale
Motor Racing at Brands Hatch in the Seventies by Chas Parker
Motor Racing at Goodwood in the Sixties by Tony Gardiner
Three Wheelers by Malcolm Bobbitt

Enthusiast's Restoration Manual Series
Citroën 2CV - How to Restore by Lindsay Porter
Classic Car Body Work - How to Restore by Martin Thaddeus
Classic Cars - How to Paint by Martin Thaddeus
Triumph TR2/3/3A - How to Restore by Roger Williams
Triumph TR4/4A - How to Restore by Roger Williams
Triumph TR5/250 & 6 - How to Restore by Roger Williams
Triumph TR7/8 - How to Restore by Roger Williams
Volkswagen Beetle - How to Restore by Jim Tyler

Essential Buyer's Guide Series
Alfa GT Buyer's Guide by Keith Booker
Alfa Romeo Giulia Spider Buyer's Guide by Keith Booker & Jim Talbott
Jaguar E-Type Buyer's Guide
Porsche 928 Buyer's Guide by David Hemmings
VW Beetle Buyer's Guide by Ken Cservenka & Richard Copping

Auto Graphics Series
Fiat & Abarth by Andrea & David Sparrow
Jaguar MkII by Andrea & David Sparrow
Lambretta LI by Andrea & David Sparrow

General
AC Two-litre Saloons & Buckland Sportscars by Leo Archibald
Alfa Romeo Berlinas (Saloons/Sedans) by John Tipler
Alfa Romeo Giulia Coupé GT & GTA by John Tipler
Alfa Tipo 33 Development, Racing & Chassis History by Ed McDonough
Anatomy of the Works Minis by Brian Moylan
Armstrong-Siddeley by Bill Smith
Autodrome by Sam Collins & Gavin Ireland
Automotive A-Z, Lane's Dictionary of Automotive Terms by Keith Lane
Automotive Mascots by David Kay & Lynda Springate
Bentley Continental, Corniche and Azure by Martin Bennett
BMC Competition Department Secrets by Stuart Turner, Peter Browning & Marcus Chambers
BMW 5-Series by Marc Cranswick
BMW Z-Cars by James Taylor
British 250cc Racing Motorcycles by Chris Pereira
British Cars, The Complete Catalogue of, 1895-1975 by Culshaw & Horrobin
Bugatti Type 40 by Barrie Price
Bugatti 46/50 Updated Edition by Barrie Price
Bugatti 57 2nd Edition by Barrie Price
Caravans, The Illustrated History 1919-1959 by Andrew Jenkinson
Caravans, The Illustrated History from 1960 by Andrew Jenkinson
Chrysler 300 - America's Most Powerful Car 2nd Edition by Robert Ackerson
Citroën DS by Malcolm Bobbitt
Cobra - The Real Thing! by Trevor Legate
Cortina - Ford's Bestseller by Graham Robson
Coventry Climax Racing Engines by Des Hammill
Daimler SP250 'Dart' by Brian Long
Datsun 240, 260 and 280Z by Brian Long
Dune Buggy Files by James Hale
Dune Buggy Handbook by James Hale
Fiat & Abarth 124 Spider & Coupé by John Tipler
Fiat & Abarth 500 & 600 2nd edition by Malcolm Bobbitt
Ford F100/F150 Pick-up 1948-1996 by Robert Ackerson
Ford F150 1997-2005 by Robert Ackerson
Ford GT40 by Trevor Legate
Ford Model Y by Sam Roberts
Funky Mopeds by Richard Skelton
Honda NSX Supercar by Brian Long
Jaguar, The Rise of by Barrie Price
Jaguar XJ-S by Brian Long
Jeep CJ by Robert Ackerson
Jeep Wrangler by Robert Ackerson
Karmann-Ghia Coupé & Convertible by Malcolm Bobbitt
Land Rover, The Half-Ton Military by Mark Cook
Lea-Francis Story, The by Barrie Price
Lexus Story, The by Brian Long
Lola - The Illustrated History (1957-1977) by John Starkey
Lola - All The Sports Racing & Single-Seater Racing Cars 1978-1997 by John Starkey
Lola T70 - The Racing History & Individual Chassis Record 3rd Edition by John Starkey
Lotus 49 by Michael Oliver
Mazda MX-5/Miata 1.6 Enthusiast's Workshop Manual by Rod Grainger & Pete Shoemark
Mazda MX-5/Miata 1.8 Enthusiast's Workshop Manual by Rod Grainger & Pete Shoemark
Mazda MX-5 (& Eunos Roadster) - The World's Favourite Sportscar by Brian Long
Mazda MX-5 Miata Roadster by Brian Long
MGA by John Price Williams
MGB & MGB GT - Expert Guide (Auto-Doc Series) by Roger Williams
Micro Caravans by Andrew Jenkinson
Mini Cooper - The Real Thing! by John Tipler
Mitsubishi Lancer Evo by Brian Long
Morgan Drivers Who's Who - 2nd International Edition by Dani Carew
Motor Racing Reflections by Anthony Carter
Motorhomes, The Illustrated History by Andrew Jenkinson
Motorsport in colour, 1950s by Martyn Wainwright
MR2 - Toyota's Mid-engined Sports Car by Brian Long
Nissan 300ZX & 350Z - The Z-Car Story by Brian Long
Pass the Driving Test by Clive Gibson & Gavin Hoole
Pontiac Firebird by Marc Cranswick
Porsche Boxster by Brian Long
Porsche 356 by Brian Long
Porsche 911 Carrera by Tony Corlett
Porsche 911R, RS & RSR, 4th Edition by John Starkey
Porsche 911 - The Definitive History 1963-1971 by Brian Long
Porsche 911 - The Definitive History 1971-1977 by Brian Long
Porsche 911 - The Definitive History 1977-1987 by Brian Long
Porsche 911 - The Definitive History 1987-1997 by Brian Long
Porsche 911 - The Definitive History 1997-2004 by Brian Long
Porsche 911SC 'Super Carrera' by Adrian Streather
Porsche 914 & 914-6 by Brian Long
Porsche 924 by Brian Long
Porsche 933 'King of Porsche' by Adrian Streather
Porsche 944 by Brian Long
RAC Rally Action! of by Tony Gardiner
Rolls-Royce Silver Shadow/Bentley T Series Corniche & Camargue Revised & Enlarged Edition by Malcolm Bobbitt
Rolls-Royce Silver Spirit, Silver Spur & Bentley Mulsanne 2nd Edition by Malcolm Bobbitt
Rolls-Royce Silver Wraith, Dawn & Cloud/Bentley MkVI, R & S Series by Martyn Nutland
RX-7 - Mazda's Rotary Engine Sportscar (updated & revised new edition) by Brian Long
Singer Story: Cars, Commercial Vehicles, Bicycles & Motorcycles by Kevin Atkinson
Subaru Impreza by Brian Long
Taxi! The Story of the 'London' Taxicab by Malcolm Bobbitt
Triumph Motorcycles and the Meriden Factory by Hughie Hancox
Triumph Speed Twin & Thunderbird Bible by Harry Woolridge
Triumph Tiger Cub Bible by Mike Estall
Triumph Trophy Bible by Harry Woolridge
Triumph TR6 by William Kimberley
Turner's Triumphs, Edward Turner & his Triumph Motorcycles by Jeff Clew
Velocette Motorcycles - MSS to Thruxton Updated & Revised Edition by Rod Burris
Volkswagen Bus or Van to Camper, How to Convert by Lindsay Porter
Volkswagens of the World by Simon Glen
VW Beetle Cabriolet by Malcolm Bobbitt
VW Beetle - The Car of the 20th Century by Richard Copping
VW Bus, Camper, Van, Pickup by Malcolm Bobbitt
VW - The air-cooled era by Richard Copping
Wonderful Wacky World of Marketingmobiles: Promotional Vehicles 1900-2000 by James Hale
Works Rally Mechanic by Brian Moylan

First published in 2003 by Veloce Publishing, 33 Trinity Street, Dorchester DT1 1TT, England. This revised second edition published in 2005. Fax 01305 268864/e-mail info@veloce.co.uk/web www.veloce.co.uk or www.velocebooks.com
ISBN 1-84584-006-2/UPC 781845-84006-8
© Des Hammill and Veloce Publishing 2003 & 2005. All rights reserved. With the exception of quoting brief passages for the purpose of review, no part of this publication may be recorded, reproduced or transmitted by any means, including photocopying, without the written permission of Veloce Publishing Ltd.
Throughout this book logos, model names and designations, etc, have been used for the purposes of identification, illustration and decoration. Such names are the property of the trademark holder as this is not an official publication.
Readers with ideas for automotive books, or books on other transport or related hobby subjects, are invited to write to the editorial director of Veloce Publishing at the above address.
British Library Cataloguing in Publication Data -
A catalogue record for this book is available from the British Library.
Typesetting (Soutane), design and page make-up all by Veloce on Apple Mac.
Printed in India

SpeedPro Series

HOW TO BUILD & POWER TUNE
HOLLEY CARBURETORS

Des Hammill

VELOCE PUBLISHING
THE PUBLISHER OF FINE AUTOMOTIVE BOOKS

Veloce SpeedPro books -

ISBN 1 903706 76 9   ISBN 1 903706 91 2   ISBN 1 903706 77 7   ISBN 1 903706 78 5   ISBN 1 901295 73 7   ISBN 1 903706 75 0

  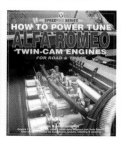 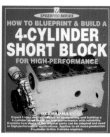  

ISBN 1 901295 62 1   ISBN 1 874105 70 7   ISBN 1 903706 60 2   ISBN 1 903706 92 0   ISBN 1 903706 94 7   ISBN 1 901295 26 5

  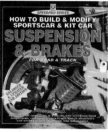   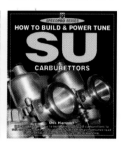

ISBN 1 901295 07 9   ISBN 1 903706 59 9   ISBN 1 903706 73 4   ISBN 1 904788 78 5   ISBN 1 901295 76 1   ISBN 1 903706 98 X

     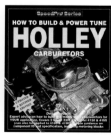

ISBN 1 903706 99 8   ISBN 1 84584 005 4   ISBN 1-904788-84-X   ISBN 1-904788-22-X   ISBN 1 903706 17 3   ISBN 1 84584 006 2

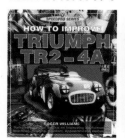  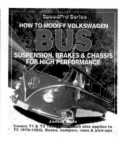 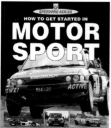 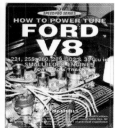

- more on the way!

ISBN 1 903706 80 7   ISBN 1 903706 68 8   ISBN 1 903706 14 9   ISBN 1 903706 70 X   ISBN 1 903706 72 6

Contents

Introduction & acknowledgements 7
Introduction 7
Bigger is not always better. 8
Acknowledgements. 9

Chapter 1. Which carburetor? . . .10
2300 11
4150 & 4160 12
Fire extinguishers: a cautionary tale. 12
Buying secondhand 13
List Numbers of common Universal . .
    Performance carburetors 14
Boosters 15

Chapter 2. Basic carburetor
identification **16**
Introduction 16
By the numbers 16
Identification examples 17
List Numbers 20
Carburetors with missing List
    Numbers 20
Holley Performance Parts Catalog. . 21
Metering blocks. 21
Has the metering block been
    modified?. 21

2300 Series carburetors 22
4150 and 4160 Series carburetors . 22
Is it a primary or secondary metering .
    block? 23
Throttle body & shaft assemblies. . . 24
Butterfly diameters in relation to
    CFM 24
CFM ratings in relation to venturi
    sizes. 24

Chapter 3. Choosing the right size
carburetor **26**
CFM calculation 27

Chapter 4. Basic carburetor
parts **29**
Repair kits 29
Finding the List Number 31
Choke 32
Fuel bowls 32
'Side hung' fuel bowls. 33
'Centre pivot' fuel bowls 36
Floats 36
Power valves 37
Accelerator pump 39
Accelerator pump cams 41

Accelerator pump discharge
    nozzles. 42
Accelerator pump mechanism fuel
    bowl non-return valves. 43
Main jets. 44
Needles and seats 46
Primary metering blocks 47
Secondary metering blocks 49
Alternative 4150 mechanical secondary
    metering blocks 49
Metering block vent baffles 50
Secondary metering plates 51
Fitting metering plates to carburetor
    bodies 54
Fuel bowl, metering block, metering
    plate gaskets 55
Vacuum secondary diaphragm
    springs. 57
Vacuum secondary diaphragms . . . 57

Chapter 5. Tuning 2300 two barrel
carburetors & the primary barrels of
4150 & 4160 four barrel
carburetors **59**
The relationship between two and four
    barrel carburetors. 59

Fuel supply 59
Introduction to tuning & adjustment procedures 61
Fuel level adjustment ('centre pivot' & 'side hung' fuel bowls) 62
Primary barrel idle speed and idle mixture 65
Secondary butterfly idle adjustment screws (4150 & 4160) 65
Holes in your butterflies? 67
Disabling the secondary barrels 68
Idle speed and idle mixture screw adjustments 69
Accelerator pump adjustment 72
Accelerator pump cams 73
Accelerator pump discharge nozzles 75
Main jets 76
Initial road or track testing to establish basic main jetting requirements . . 77
Why engines fitted with long duration camshafts need low rated power valves 78

Fuel bowl leaks 78
Summary 78
Further tuning of four barrel carburetors 79
Specific information on 2300 two barrel Holley carburetors 79
Accelerator pumps 81
Main jets 81
Power valves shutting at high rpm . . 81
Power valve selection 84
Four barrel to two barrel option . . . 86

Chapter 6. Tuning the secondary barrels of 4150 & 4160 carburetors 88
Using acceleration timing to optimise jetting 89
Tuning vacuum secondary carburetor secondary barrels 92
Changing main jet sizes (metering plate carburetors) 92
Changing main jet sizes (metering block carburetors) 92

Power valves 92
Selecting vacuum secondary diaphragm springs 93
Tuning 'double pumper' or mechanical carburetor secondary barrels . . . 93
Secondary idling system 94
'Four corner' idle mixture adjustment 94
Accelerator pump settings 94
Optimising main jets 96

Chapter 7. Inlet manifolds 97

Chapter 8. Aftermarket carburetor components 101
Weber carburetor air/fuel metering for Holley carburetors 101
Secondary metering plates 102

Appendix - Holley carburetor numerical listings 104

Index .127

Introduction & acknowledgements

INTRODUCTION

This book specifically covers Holley 'Universal Performance' and 'Competition' two and four barrel carburetors. These carburetors have been made in huge numbers since production of four barrel carburetors for Ford started in 1957. Included in this book's coverage are 2300 series 350 and 500-CFM two barrel, 4160 series 390 to 780-CFM vacuum secondary four barrel and 4150 series 390 to 1000-CFM 'double pumper' four barrel carburetors. The carburetors covered are the most common models, those that readers can easily buy new or secondhand.

Even within the confines of the carburetor types covered by this book, the range of Holley carburetors is vast when specifications for different original equipment applications are taken into account, this is because of the sheer number of derivatives that Holley made for various USA auto manufacturers. Many of these individual carburetor models are now obsolete, but parts are still available new or secondhand.

Note that the acronym 'CFM' is used constantly throughout this book and is an abbreviation of 'Cubic Feet per Minute' which is the amount of air individual carburetors can flow as rated by Holley using the company's own criteria. Two and four barrel carburetors are also rated differently.

Because of the difficulty many people have with tuning these excellent carburetors, this book has been structured to lead readers into and through various procedures in a logical manner. This approach creates some degree of repetition, of course, because of the need to expand explanations of some aspects of how these carburetors work and how they need to be tuned. Understanding the workings and tuning procedures of these carburetors will require some effort to grasp, but the effort will be very worthwhile.

A vitally important aspect of this book's coverage is detailed guidance on how to: (a) choose the right size, model and specification of Holley carburetor for any individual high performance application, and (b) set up and tune that carburetor for optimum performance with minimum difficulty. The necessary calculations for optimum carburetor selection are detailed, as is an all important - and realistic - efficiency rating guide.

The tuning techniques described in this book are specifically designed for the carburetors already mentioned, but can also be applied to most other four barrel Holley carburetors. The techniques are designed to prevent readers from making the common mistakes that lead to frustration and poor engine performance, especially with the four barrel carburetors. In spite of the length of time that Holley carburetors have been available, the same mistakes in tuning, due to misleading information and misunderstandings, are made over and over again by a huge number of people ...

Following the tuning guidelines detailed in this book will reduce the likelihood of the sort of errors which can lead to these excellent carburetors giving less than satisfactory performance. Poor performance almost always stems from a poor choice of carburetor size (CFM), a wrong part or parts, a damaged carburetor or an incorrect setting or settings.

This book does not cover modification of these carburetors (except for one instance intended for use if all else fails) beyond optimising component choice and jetting. So, in broad principle, this book is about getting what you have got right once you've chosen the correct size and type of carburetor for your application.

If you're going to buy a new carburetor, it's better to buy a carburetor which has exactly the right specification for your application as it comes from Holley. If you choose wisely, you'll never have to change the carburetor's specification and it will be precisely what you want from day one. A similar looking but less expensive new carburetor may not be a bargain for, while all Holley carburetors are good, some are more adjustable and have better features. Having to buy alternative parts to uprate a poorly specified carburetor is almost always an expensive thing to do. However, there is no point in spending money on features that you definitely do not need for your particular application.

Over the years there have been so many high performance derivatives of Holley carburetors that it can be quite confusing sorting out just what parts can go with which carburetor. That said, many Holley carburetors can be converted into high performance units, or better high performance units, by fitting alternative parts. This 'mixing and matching' approach can be very cost effective if you use carefully chosen secondhand components. Taking advantage of the Holley's modular construction can be useful if you already have a carburetor but need to upgrade it to make it more suitable for your application. For

> **Bigger is not always better**
> One race day a race car owner/driver came over to look at the author's sportscar's 1962 221 cubic inch smallblock Ford V8 engine (a unit hugely criticised by the US motoring press of the day as being 'useless') with its 500CFM two barrel Holley fitted to a four barrel Torker 289 inlet manifold. He asked: 'how can an engine go so well with such a small carburetor?' I simply told him that, as far as I was concerned, it was big enough for the particular engine and that I'd tried a 'correctly' sized vacuum secondary four barrel carburetor and the car had not gone any better. In fact I didn't think it went as well overall. I also said that I liked the simplicity of the two barrel carb and, while it mightn't look the business, at least it was well tuned and never caused me any bother at all.
>
> Some months later I trial fitted a 750CFM four barrel but blanking off a few things and only using the two front chokes (i.e. equal to the 500CFM two barrel). The engine definitely looked much more impressive ... but went exactly the same, at 11.9 seconds for the 1/4 mile!
>
> My car was not the fastest car around by a long shot, but it was 100% reliable and always went well, which was more than could be said for about 50% or more of the other competitors cars which, for a variety of reasons (none of them good), always seemed to have some problem or another such as misfires and so on. Once the carburetor I was using had been tuned, it didn't need to be touched again except for routine maintenance and I didn't even have to do much of that.
>
> On another occasion a racing car owner came over to me and asked if I'd look at his car's engine and, perhaps, be able to tell him why his car never seemed to go well in spite of the large amount of money he'd spent on it. I took one look at his fantastic looking V8 engine and told him to take the whole induction system off, fit only one of the two carburetors to a 360 degree single four barrel aluminium aftermarket inlet manifold and try that.
>
> Shocked by what I'd suggested, and clearly harbouring grave doubts, he came around to the idea after a friend offered to lend him a suitable 360 degree inlet manifold for a while. Afterwards the difference in performance was immediate and very much for the better, although the engine didn't look as impressive as it once had.
>
> Looks can be deceiving and complication can lead to tuning problems that we can all do without. While two four barrel carburetors on a tunnel ram inlet manifold might look the business and might work exceptionally well in certain instances, most enthusiasts should stay well away from this sort of gear. Keeping everything as simple as possible is a well founded principle for most of us, as is the developmental approach of getting the best out of what you've already got before considering progressing to more complicated systems.

instance, you might want to fit 'centre pivot' instead of 'side hung' fuel bowls.

Though the three series of Holley carburetors covered by this book are still being manufactured in various CFM sizes, many of the Holley carburetors in use now, or available as secondhand units or spares, are of considerable age with an unknown history. This can result in various problems, such as wrong - but not obviously so - combinations of components. Identification details given in this book should allow you to ascertain the true specification, and therefore suitability, of all components.

Using modified and, as a consequence, perhaps non-standard sized components can be the cause of endless tuning problems. Careful scrutiny of all of the carburetor parts usually reveals which components have been tampered with. When drillings are drilled out to larger sizes, they're never drilled as neatly as in original manufacture and this is a definite giveaway. There will likely be a burr thrown up on a hole that has been drilled after original manufacture and the yellow chromate surface coating put on the carburetor parts by Holley will be removed meaning that there will be a color difference where a drill has been used.

As has already been implied, one of the aims of this book is to keep everything as logical and as simple as possible so that anyone with a modicum of mechanical understanding can pick up the basics of setting up and tuning these carburetors and work out solutions for themselves. There's nothing more frustrating than spending a lot of time and money on carburetors and inlet manifolds, for example, which you think should work, but don't. When this happens it's not time to panic and think about getting rid of the whole set-up, it's time to go back to square one and systematically go through the specification again. You don't have to be a 'tuning wizard' to get a Holley carburetor tuned correctly. There is no magic involved here in getting the best results, just clear thought, a plan of action and some small amount of work.

There isn't a replacement part or tuning part for these Holley carburetors that cannot be bought in any developed country in the world. There are also many aftermarket manufacturers which make top quality replacement parts and specialist performance parts that offer some improvement. These aftermarket parts are often as widely available as genuine Holley parts.

ACKNOWLEDGEMENTS

I would like to acknowledge the valued assistance given to me by Mr. Thomas R. Kise of Holley's technical helpline <help@support.holley.com> in answering all of my queries.

Many thanks to the following for the lending me Holley carburetors and parts for photographic purposes and general use during the writing of this book: Bob Aspinal of Rodley Motors, Bradford, England, Ian Richardson of Wildcat Engineering, Rhydymain, Gwynedd, North Wales, Brian Wills of Kings Mews Racing, Newton Abbot, England, to Paul Kynaston of Kynaston Auto Services, Marsh Barton, Exeter, England and to Barry Dufty of Specialist Auto Parts, Yeovil, England; Dave Mills of Road and Track Performance, 159 Manukau Road, Pukekohe, New Zealand; George Sheweiry, Green Bay, Auckland, New Zealand; Dwight and Ron of Performance Parts, Papakura, New Zealand; Etchells Race Parts, Tauranga, New Zealand; and last but not least, thanks to my wife Alison for the construction and maintenance of my website www.deshammill.co.uk and the assistance she gives with my books.

www.velocebooks.com/www.veloce.co.uk
All books in print • New books • Special offers • Newsletter

Chapter 1

Which carburetor?

The carburetors detailed in this chapter are the best Holley carburetors to buy, new or secondhand, and to use in most performance applications. This is by virtue of the fact that they come as standard with all of the right components. They can be tuned with a minimum of fuss, and this is more than can be said for many other engine specific Holley derivatives which can be more trouble than they are worth when used in different applications than those for which they were intended. The latter is no reflection on Holley, which makes carburetors for many applications, and the catalogs clearly state which carburetors are Universal Performance ones and which are for high-performance applications. The recommendation is to get the right model of the carburetor for your application in the first place and work from there. Doing this will save time and money and, almost certainly, you'll end up with a better result. For example, if your application is one

Four barrel Holley carburetors and aftermarket inlet manifolds, as seen on this 351ci Cleveland engine fitted into an early 1970s Australian Falcon, are still very popular for racing purposes. They are comparatively simple, easy to maintain, and deliver excellent torque and power for a very reasonable price.

WHICH CARBURETOR?

Triple two barrel carburetor set-up. This was a popular original equipment induction system for big block engines in the 1960s.

which is going to need 'centre pivot' fuel bowls you might as well have them from day one as opposed to trying to use 'side hung' fuel bowls only to have to change them at a later date.

The Holley range of 'Universal Performance carburetors' and 'competition carburetors' have been clearly listed in Holley literature for years. What is not always so clear, however, is the ideal application for each derivative of carburetor. Knowing the limitations of each model of carburetor is vital, and it's better to know the limitations before you start spending money ...

The 2300 series two barrel carburetors, at 350 and 500-CFM, are mechanical in operation and, when ordered in 'Universal Performance' form, are more or less race ready out of the box. With 'centre pivot' fuel bowls and a mechanical choke, these are the best models to have.

There are several other derivatives of the 2300 which have 'side hung' fuel bowls and automatic chokes. Although very good carburetors for some applications, they are more limited by their specification. These carburetors can, of course, be upgraded to higher specification.

Some models of these 2300 two barrel carburetors are not suitable for single carburetor applications. They were originally intended to be used in three carburettor applications with a progressive throttle linkage (used in this way by Chevrolet, Ford and Chrysler on their big block engines as standard equipment). Only the centre carburetor of the original three can be used as a standalone carburetor.

2300

The 'Universal Performance' 2300 carburetors and the 'Competition' 2300 carburetors have a primary metering block with idling circuits, a power valve and an accelerator pump mechanism. These carburetors can be used for all road going applications and, because these versions are all equipped with centre pivot fuel bowls, they can be used for all racing purposes too. On the basis of simplicity versus potential performance it doesn't get much better than this, provided the carburetor is large enough for the engine. Being a two barrel unit there are fewer tuning options and complications.

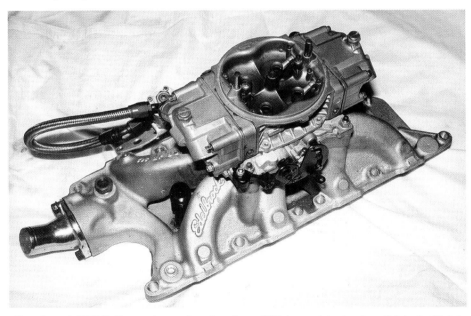

Four barrel 4150 Holley racing carburetor atop a 360 degree/single plane 'high rise' inlet manifold.

SPEEDPRO SERIES

This throttle body and shaft assembly had no numbers stamped on it at all, but the butterfly bores measured 1.680in (1¹¹⁄₁₆in) in diameter making it suitable for this carburetor when checked against the Numerical Listing specified sizes. A vernier calliper is shown being used to measure the throttle bore (the throttle body and shaft assembly has been removed from the carburetor body in this photo).

4150 & 4160

Next come the 4150 and 4160 four barrel Holley carburetors. The primary side of these four barrel carburetors is identical for both and to the two barrel 2300 series of carburetors in virtually every physical aspect and working principle. The vacuum secondary and mechanical secondary 4160 and 4150 four barrel carburetors all have primary metering blocks with adjustable idle circuitry, a power valve, and all have a single primary accelerator pump mechanism (less common 'centre squirter' excepted).

What defines a Holley four barrel carburetor as a 4150 or a 4160 is the fact that 4150 carburetors have a secondary metering block fitted as standard and 4160 carburetors have a secondary metering plate fitted as standard. Further to this, most 4160 Holley carburetors can be converted into 4150 carburetors by fitting a kit that you can buy from Holley or build yourself from parts from suitable secondhand carburetors. The main advantage of converting a 4160 into a 4150 carburetor being that secondary jetting changes are easily made compared to having to get alternative metering plates or drill out the existing plate to make changes. Also, once the metering plate is drilled out to a size that is too large, it can't easily be changed back to what it was.

The differences that need to be taken into account between four barrel carburetors are essentially to do with the secondary side. An example of this is when, on a road car, a 600-CFM 4160 List Number 1850 vacuum secondary four barrel carburetor is replaced by a 600-CFM 4150 List Number 4776 'double pumper' four barrel carburetor. The engine will not usually go any better anywhere in the rpm range in spite of the 4776 carburetor having some very desirable

Fire extinguishers: a cautionary tale ...

The following incident, as witnessed by the author, serves to illustrate how events can get out of hand, and how having an extinguisher could make a big difference.

"Whilst walking down a street somewhere in the USA, I heard an engine being turned over, and over, and over. Then, suddenly, there was a very loud backfire and a *whooooosh* noise. I stopped and looked around to see if I could locate the noise: it was a small pickup truck with its hood up and with the air cleaner sitting on the ground next to it. The driver had just got out of the cab and was at the front of the pickup looking at a few small flames on the top of the engine. He casually walked to the back of the truck and came back with a sack and proceeded to beat away at the flames. Well, the sack caught fire, the underside of the hood caught fire and flames started coming out of the grill; another *whooooosh* and the whole engine compartment was a mass of flames and, finally, the driver caught fire too. He quickly moved away from the pickup to put out the flames on his clothing with his bare hands, luckily he received no serious burns. In no time at all the pickup was burning from end to end. The fire brigade arrived and put the fire out but, by which time there really wasn't much of the pickup left - except the air cleaner which had been sitting on the ground away from the action.

"After the smoke had wafted away, I ventured over to the driver and asked him why he'd beaten away at the flames instead of just trying to smother them? The stunned driver, scratching his head, turned to me and said, "beating away at the flames worked fine all last week!"

WHICH CARBURETOR?

tuning features. Retaining the vacuum secondary 1850 usually results in a more driveable road car and at lower cost ...

BUYING SECONDHAND

Buying a secondhand Holley carburetor can be a risky business, especially if the seller is deliberately trying to deceive you by trying to sell you something damaged, incomplete or a collection of unmatched parts. Never pay full value for a secondhand carburetor unless you are able to positively identify all of the main parts and ascertain the overall condition of the carburetor is good. This vigilance may still not save you from making a mistake as parts can be drilled out, modified or even be missing altogether without it being very obvious.
The huge importance of correct specification is the reason this book is 'loaded' with identification material.

If a secondhand carburetor looks like new the chances are that it is relatively little used and the seller will usually want a high price for it. The difference between a near new carburetor and a cleaned up carburetor is usually apparent. It's the older, dirty looking, carburetor that may, or may not, be a collection of mixed up parts that usually needs the most careful scrutiny. Nevertheless, after a thorough clean up, what was a dirty old carburetor can turn out to look like a near new carburetor and, after jetting it correctly for the application, can prove to be an unbelievable bargain and give absolutely troublefree service.

Holley carburetors are pretty robust, with many of those in regular use today being more than four decades old. They'll survive all manner of use and abuse, and after a proper rebuild, are seemingly able to go through it all over again. Condition is more important than age when it comes to secondhand carburetors, with wear of (and fuel leakage from) the primary spindle being a common indicator of general wear in what otherwise looks to be a carburetor in very good condition. Another indicator of excessive wear is loose staking on the end of the primary spindle where the throttle linkage arm is positioned. This staking frequently works loose after a lot of work (but can be silver soldered to restore to as new). A loose secondary actuating linkage is another indicator of excessive wear, resulting in the secondary butterflies not shutting off correctly, for example. Another problem that is found from time to time is butterflies that are not at 90 degrees to the throttle bore at full throttle. The throttle stop is the usual problem and it will often have to be hand filed (slightly).

Many owners of high performance cars have never bought a brand new Holley carburetor in their lives, and have no intention of ever doing so as new Holley carburetors are relatively expensive. What such people frequently do is to have a few secondhand models of the carburetor type that they use so that they have a selection of spare parts of the right type. It has not been uncommon for folks to start off with, say, a 600-CFM vacuum secondary carburetor which has 'side hung' fuel bowls and a metering plate and then uprate it by fitting 'centre pivot' fuel bowls and a 4100 rear metering block, and so on, and ending up with a more suitable carburetor for their application at a very low cost. However, you can only do this if you know the true sequence of the modular nature of these Holley carburetors: then it's quite possible to build up an excellent carburetor from a pile of seemingly miscellaneous bits and pieces. The secret to doing

Inside callipers can be used to measure the size of the primary and secondary venturis. It can be a bit difficult to measure the venturis of the smaller CFM carburetors, such as the 390-CFM four barrel for instance, in this way without removing the throttle body and shaft assembly and checking the sizes of the venturis from the butterfly side of the main body. A Vernier calliper is used to measure the distance between the tips of the inside calliper to obtain the diameter.

SPEEDPRO SERIES

this is to be able to positively identify all of the secondhand parts you are going to use and be sure that they have not been tampered with and that all passageways are clear and able to work. One set of holes that always needs to be checked and cleared of blockage, for example, is the secondary idle passageways in the four barrel carburetor throttle body and shaft assemblies. These holes are very small and do get blocked up with use. They are found directly below the progression slots.

In some instances on secondhand carburetors, one or both metering blocks may have been swapped for metering blocks that are from different models of carburetor. They might work on the 'wrong' carburetor or they might not! If the metering blocks are from another carburetor with the same CFM rating and have not been tampered with, the chances are that they will work well.

Spindle wear can be a vexing problem, and even an apparently new carburetor can have a spindle fitted to it that is really far too loose. Even when buying a new Holley carburetor, always check to see how much up and down movement there is in both primary and secondary spindles. Don't buy it if there is excessive play in the spindle, or spindles, as fuel will start to leak out between the spindle and throttle plate hole much earlier than it normally would.

Check all spindles for wear, especially the primary barrel spindle on four barrel carburetors as these get more use than the secondary spindle on a four barrel carburetor. This is one aspect of Holley carburetors that can be the cause of major fuel leakage/seepage and there is nothing that can be done about it except get a new throttle body or have the old throttle body repaired. Fuel stain marks around the primary spindle are a real giveaway to this condition, especially on the side of the carburetor to which the throttle linkage connects.

Warning! - Take no risks with fuel leakage from the throttle spindle as an engine fire could result. Also, anyone who intends to tune cars and, more specifically, work on carburetors and fuel systems should have a good-sized fire extinguisher appropriate for fuel fires on hand at all times. Such extinguishers are inexpensive and readily available from all automotive parts suppliers.

While a worn or loose spindle problem can be effectively repaired, it does mean that the spindle will have to be removed from the carburetor's throttle plate and this is the problem. Some people are very good at repairing tricky carburetor problems like this for a realistic price, and the finished article can be as good as new. Check the internet for a Holley carburetor repair specialist. The major problem with doing this work is actually getting the spindle out of the throttle body without damaging the spindle or butterflies. Holley doesn't want the screws that secure the butterflies to the spindle to come out in service so the ends are riveted after fitting. It is not impossible to remove the screws but it is quite difficult unless you have the right sort of equipment. Specialists who repair carburetor throttle body and shaft assemblies have all of the right equipment and new parts on hand. Pay to have this exacting work done for you, it's simply not worth trying to do it yourself.

Later Holley carburetors have replaceable Teflon bushings, as opposed to the spindle working directly in the cast aluminum baseplate. The spindles and butterflies still have to be removed to replace the Teflon bushes which still means it's specialist work.

These carburetors are not particularly noted for their economy when producing maximum power in a racing situation. However, in a road going application they can be set up to give very reasonable economy in relation to the power being produced and they certainly can equal similar CFM sized carburetors from other manufacturers if correctly set up.

No Holley carburetor will give good fuel economy if it's too large for the particular engine it's being used with. For a start, the accelerator pump shot will usually be larger than necessary if the primary two barrels are a lot larger than they need to be. This usually means that a large portion of the partially burnt pump shot is exhausted by the engine as a puff of black smoke. This is wasteful and serves no useful purpose whatsoever. If the carburetor barrels are ideally sized, fuel economy will be good without any loss of engine performance. This factor may, in fact, mean fitting a smaller carburetor to your engine than you think it should have.

LIST NUMBERS OF COMMON UNIVERSAL PERFORMANCE CARBURETORS

2300 - 350-CFM 2 barrel - List 7448
2300 - 500-CFM 2 barrel - List 4412

Mechanical secondary carburetors (all have metering blocks)
4150 - 600-CFM 4 barrel - List 4778
4150 - 650-CFM 4 barrel - List 4777
4150 - 700-CFM 4 barrel - List 4778
4150 - 750-CFM 4 barrel - List 4779
4150 - 800-CFM 4 barrel - List 4780
4150 - 850-CFM 4 barrel - List 4781
4150 - 950-CFM 4 barrel - List 8049
4150 - 1000-CFM 4 barrel - List 8051

Vacuum secondary carburetors (with metering blocks)

WHICH CARBURETOR?

4150 - 600-CFM 4 barrel - List 4742
4150 - 600-CFM 4 barrel - List 2818
4150 - 600-CFM 4 barrel - List 80145
4150 - 725-CFM 4 barrel - List 4118
4150 - 780-CFM 4 barrel - List 9188
4150 - 855-CFM 4 barrel - List 3418

Some of these carburetors are better to use than others in certain circumstances. For example, the 4742 can be better than the 2818 because it comes as standard with centre pivot float bowls. This is particularly relevant if you are into circuit racing.

Vacuum secondary carburetors (with metering plates)
4160 - 390-CFM 4 barrel - List 6299
4160 - 390-CFM 4 barrel - List 8007
4160 - 450-CFM 4 barrel - List 4548
4160 - 600-CFM 4 barrel - List 1850
4160 - 600-CFM 4 barrel - List 9834
4160 - 750-CFM 4 barrel - List 3310
4160 - 780-CFM 4 barrel - List 7010

Some of these carburetors are better to use than others in certain circumstances. For example, the 6299 can be better than the 8007 as it comes with a manual choke as standard (as opposed to an automatic choke). This is, of course, a relatively minor consideration.
Reminder! - A 4150 four barrel

This Holley carburetor has 'dog leg' boosters.

carburetor has a metering block and a 4160 a metering plate. However, it is quite possible to make a 4160 carburetor into a 4150 using a Holley conversion kit (or adapting secondhand components) so that the rear jets are replaceable.

BOOSTERS

There are subtle differences between many of the Holley carburetors but one quite outstanding, though not particularly well understood one, is the fact that there is a difference between the discharge boosters that you need to know about. There is a 'dog leg' booster and a 'straight leg' booster. The 'dog leg' booster is the one that a competition engine's carburetor needs to have. This may

This Holley carburetor has 'straight leg' boosters.

limit the carburetor CFM sizing range that can be used, but engines almost always accelerate better with this type of discharge booster, even if the carburetor is a fraction too large for the engine. Most vacuum secondary four barrel carburetors have 'straight leg' boosters, but all mechanical secondary four barrel carburetors in the range of carburetors suitable for these engines have 'dog leg' boosters, as do 500CFM two barrel carburetors. This single advantage is one of the reasons why the 715CFM List number 3259 vacuum secondary carburetor was used by Shelby American in the 1960s rather than a smaller carburetor.

www.velocebooks.com/www.veloce.co.uk
All books in print • New books • Special offers • Newsletter

Chapter 2
Basic carburetor identification

INTRODUCTION

Holley's website (www.holley.com) is packed with information, including lists of carburetor models with individual specification information and much more. The lists and specifications can be downloaded in pdf file format which means you can keep copies on your computer or even print them.

A wide range of useful parts information is also listed in Holley's annually published *Holley Performance Parts Catalog*. Anyone who is using and tuning 2300, 4150 or 4160 carburetors will find this catalog useful. There is no one catalog available from Holley which, realistically, gives all the information you will need especially if you are dealing with a secondhand carburetor which has had parts swapped around at some stage.

By the numbers

However, the information on the website and in various Holley publications, does not tell you ALL you need to know, especially if you're dealing with secondhand carburetors and trying to identify the specification of certain parts. For instance, though the Numerical Listing on the Holley website lists the primary and secondary metering blocks for individual carburetor models by part number (which is good if you are intending to buy new metering blocks), this information is of no use at all if you are trying to identify the metering block(s) that you have in your possession. The part number and the component identification numbers stamped on the metering blocks bear no relation to each other whatsoever. It's alright if the List Number happens to be stamped on the metering blocks, but most carburetors are not like this. This is a major problem for owners of secondhand carburetors and the only thing to do is go back to Holley and ask what numbers were stamped on the metering block(s) of the particular List Number of carburetor. If the numbers are different, the metering blocks are not original and, unless the Holley Technical Services representative you get in touch with recognizes the numbers you send, he or she won't be able to tell you what carburetor the metering blocks are originally from. This problem illustrates the reality of the situation.

The modular nature of these carburetors, while being an excellent feature, can be a problem too. It's amazing how many used carburetors end up with the wrong components fitted and this can really matter when it comes to metering blocks, for example. Uniquely, this book allows the identification of the specification of all significant components of the subject Holley carburetors so that anyone who has a used Holley carburetor will be able to ascertain whether the particular components - including metering blocks - are compatible or not. This feature is of major significance because

BASIC CARBURETOR IDENTIFICATION

Identification examples

One of the main problems with secondhand Holley carburetors or carburetor components of unknown origin/history is positively identifying (and therefore knowing the specification of) the various parts, such as the metering blocks, carburetor bodies, throttle bodies and shaft assemblies. It is definitely advisable to have identified all of the major components before buying a secondhand Holley carburetor.

During the original research for this book the author was checking the identification of a used four barrel vacuum secondary carburetor. The number '3259-1' was found stamped on it and checking the current Numerical Listing on Holley's website identified the carburetor as a 725-CFM 4150 model.

With the carburetor fully stripped the author could see that the body of the carburetor could also take a rear mounted power valve (as it wasn't flat backed), but that the body had not been drilled to make such a valve operative. This means that the carburetor was not originally intended to have an operative power valve for its original equipment application. The throttle body and shaft assembly had been drilled, but not the main body.

There were no numbers of any description stamped on the throttle body and shaft assembly. The author measured the throttle bore diameters with a vernier calliper to make sure that the butterflies were the right size for this model of carburetor (these sizes are given in Holley's numerical listings. **See picture below.**

What could not be ascertained without the help of Holley's Technical Helpline was whether the two metering blocks were the right ones for the carburetor or not. The numerical listings did not give any indication of the specification of the primary and secondary metering blocks though they did show that both the primary and secondary metering blocks are 'N/S' (non serviceable) items, so you can't buy new metering blocks for this particular carburetor.

E-mailing Holley and quoting the number 3259 and all of the number groupings on the metering blocks resulted in the following information. The particular carburetor was a 725-CFM vacuum secondary 4150 model carburetor which had originally been fitted to a 1965 Cobra Mustang with a High-Performance 289 engine. As the throttle body and shaft assembly had no numbers Holley couldn't help with identity, but the butterfly diameter sizes were the right sizes for the particular carburetor. From the numbers given, Holley confirmed that the secondary metering block was the correct one, but the primary metering block was incorrect. The numbers for the correct metering block were given in the reply from Holley. Although unable to say what carburetor model the primary metering block I had was from, the reply did say that I could use a primary metering block from another similar CFM rated four barrel Holley carburetor (700 or 750-CFM).

Another carburetor of unknown origin that came the author's way was a 1850-2 unit which the Numerical Listing says should have a throttle body and shaft assembly numbered 8382, a primary metering block with 8485, 1947 and 3 stamped on it, and a secondary metering block with 4241, 4100 and 3 stamped on it. In fact all of the main components had a different series of numbers to the List Numbers quoted by Holley ...

After e-mailing Holley with all of the numbers, the reply stated that the parts were in fact the originals, with the exception of the secondary metering block (which was the unit used by Holley in the kit 34-6 to convert a 4160 carburetor into a 4150 carburetor). All parts were, as a consequence, correct.

To give me this sort of information the technical staff at Holley have to draw the BOM or 'build of material' sheet that they have for each individual carburetor Holley has ever made to find out what was originally fitted to a carburetor. All specifications of the carburetor build are on these sheets and only on these sheets.

there are millions of Holley carburetors out there, all with numbers stamped on their individual components which are near meaningless unless you work for Holley and have direct access to their vast product information database.

Holley offers technical help via e-mail <help@support.holley.com> which can be used by anyone to ask direct technical questions. When you send an e-mail initially you get an automated response irrespective of where you are in the world. The reply will usually follow within one, perhaps two, working days which is very reasonable. The technical staff are helpful and the identification information about any particular carburetor absolutely accurate.

The List Number as stamped on the carburetor body (choke tower usually) is the key to identifying a carburetor as the (original) specification can be derived from it. If you give the List Number to Holley, the technical staff can identify the carburetor, irrespective of when it was made.

When you make an enquiry to Holley Technical Services by e-mailing a List Number and asking for the metering block identification numbers, they draw the relevant 'build of material' (BOM) sheet to get precise information for you. Holley has well over 100 filing cabinets containing thousands of BOM sheets. All of the numbers stamped on the other components of a carburetor will be listed on the BOM sheet (that's the numbers on the throttle body and shaft assembly, the primary metering block and the secondary metering block, for example). These other parts might have the List Number stamped on them as well as individual part numbers. This inconsistent numbering can be confusing when some carburetors quite clearly have the List Number stamped on the body of the carburetor, the primary metering block, the secondary metering block and the throttle body and shaft assembly, while other carburetors do not.

The two barrel Holley carburetor List Number is found on the choke tower. Relevant numbers on this particular carb are 'LIST-4412-S' with '4412' being significant numbers making this particular carburetor a 500CFM one. The numbers 2572 underneath are not relevant to the identification of the carburetor, just the top line after the word 'LIST.'

The throttle body and shaft assembly of this two barrel carburetor has the numbers '4412' stamped on the top right-hand side, adjacent to where the metering block and the fuel bowl fit. This unit is from a 500-CFM two barrel carburetor.

When individual Holley components are made they're stamped with letter/number combinations or just numbers to aid positive identification.

BASIC CARBURETOR IDENTIFICATION

The four barrel carburetor List Number is found on the choke tower. The relevant numbers here are 'List-8007' with the '8007' identifying this particular carburetor as a 390CFM one. The numbers '1780' underneath are not relevant to the identification of the carburetor.

The metering block has the numbers '4412' stamped on the top side and, in this case, on the right-hand side. This unit is from a 500-CFM two barrel carburetor.

The four barrel carburetor throttle body and shaft assembly has the List Numbers stamped on the underside of it. This particular throttle body and shaft assembly has '1850 2' stamped on it, '1850' being the significant numbers. This unit is from a 600CFM vacuum secondary carburetor.

This numbering code system is perfect for manufacturing and replacement parts supply, but is not so good for end users dealing with secondhand carburetors which could have had wrong parts installed at some stage, which is easy to do with the modular 2300, 4150 and 4160.

While a complete part number code listing would be nice, one isn't available from Holley and never has been. However, the next best thing is getting in touch with Holley with the List Number and the numbers of the individual components you want identified. This doesn't mean that Holley will necessarily be able to tell you which particular carburetor the parts are from if they're not part of the original specification of your carburetor. There is no published cross reference of metering blocks and carburetors. This makes the identification of the millions of metering blocks made by Holley over the years very difficult - but not impossible with this book's help - to categorise for further use.

This business of correctly identifying secondhand carburetors or individual secondhand parts, such as metering blocks, has led to massive confusion and, over the years, has resulted in millions of Holley carburetors and parts being discarded. Holley carburetor production peaked at 1 million units per month some years ago, so the numbers of carburetors and individual components available today is staggering. I can't begin to guess the total number of times I've been into garages, engine reconditioning workshops, engineering shops, private garages, race shops, engine component supply shops and seen a wide variety of Holley carburetor components sitting on shelves or bench tops and been asked, "What type of Holley carburetors are these parts from?" In nearly all

SPEEDPRO SERIES

instances the answer has been "I don't know", which was usually followed by, "If you want them you can have them, they've been here for years and we don't have a use for them". Complete carburetors in perfect and brand new condition have come my way on many occasions so, consequently, I have a large range of carburetors and parts which have come in very handy over the years. However, it has only been while preparing this book that I have really delved deeply into Holley component identification which has proved a real experience, but not one I'd wish to repeat! The result of all this research is, as far as it is possible to do so, the solution to secondhand carburetor and individual part identification. The only components that cannot be 100% accurately identified will be some metering blocks, but ALL metering blocks can be categorized to the degree that you'll know whether or not a particular metering block is suitable for use on your carburetor and likely to work well. This sort of information is unique to this book.

List Numbers

The bodies of all of the Holley carburetors covered by this book have a List Number stamped on the choke tower. Every aspect of original carburetor specification and application can be ascertained from the List Number, making this the most important series of numbers stamped on a Holley carburetor. The accompanying photos show the basic positions of identification numbers.

The List Number 8007 identifies this carburetor as a 390-CFM four barrel. The numbers 7800 3 are stamped on the underside of the throttle body and shaft assembly and the numbers 77973 on the top side of it. The numbers 8909 are

Inside callipers can be used to measure the size of the venturis.

stamped on the top left-hand side of the primary metering block and the numbers 7204 and 5 are stamped on the right-hand side of the primary metering block. The metering plate has 34 stamped in the middle of it. Visual inspection reveals that this carburetor is a vacuum secondary 4160 because it has a vacuum diaphragm housing and a metering plate. It also has an automatic choke fitted. Measuring the butterfly diameters confirms that the throttle body and shaft assembly is for a 390-CFM carburetor without the need to check with Holley's technical department. The only major part of this carburetor that is not easy to identify by the numbers stamped on it is the metering block, but in this case there are two other clues as to why the correct primary metering block is fitted. The first is that it had 0.039inch diameter power valve restriction channel (PVCR) holes drilled in it, and the second is that it has an accelerator pump transfer tube fitted which matches the front of the body of this model of four barrel carburetor.

Carburetors with missing List Numbers

If you ever find a carburetor which has had its choke tower machined off at some stage, and therefore had the List Number removed, all is not lost. Carburetor venturi sizes are listed in the Numerical Listing and the actual venturis can be measured with inside callipers. Venturi sizes can then be compared to those detailed in the Numerical Listing and the approximate CFM rating of the carburetor body deduced. Whether or not the carburetor has mechanical secondary or vacuum secondary can be ascertained by visual inspection. The metering block numbers may also be of help, as might the throttle body and shaft assembly as they might well have the List Numbers stamped on them. There is quite a bit you can do to identify a Holley carburetor body in this condition. In fact, there is quite a lot you can do to identify virtually any Holley 2300, 4150 and 4160 carburetor enough to categorize all of the vital components and rebuild it and

BASIC CARBURETOR IDENTIFICATION

re-use it knowing that everything will work as it should.

HOLLEY PERFORMANCE PARTS CATALOG

Holley Performance Parts Catalog Numerical Listing is a very comprehensive listing, although it doesn't give all of the information you might require. Other parts listings in the *Holley Performance Parts Catalog* fill in most of the detail. Within the catalog you'll find listings for secondary metering plates, secondary diaphragms for vacuum secondary carburetors, power valves, secondary main jets as fitted to metering blocks, accelerator pump discharge nozzles, standard main jets (with jet diameter sizes in thousandths of an inch) against the various numbers as stamped on the jet, needle and seat assemblies and floats. The Numerical Listing gives details of throttle bore diameters and the venturi sizes for various CFM ratings. The catalog also lists metering blocks by part number for the ordering of new ones from Holley, but not stamped on component identification numbers.

Note that the *Holley Performance Parts Catalog* uses the abbreviations 'N/S,' 'N/A' and N/R.' There's no key or mention of what they stand for in the catalog, however N/A is a universal abbreviation meaning 'not applicable,' N/R stands for 'not required' (on the particular carburetor) and N/S means 'non serviced' (there is no sales number for this part and you cannot buy it as a spare part). There is mention of these abbreviations on Holley's website, but only at the very end and well beyond the last specification listing.

METERING BLOCKS

Metering blocks have always been difficult to identify if the numbers stamped on them do not match the

Has the metering block been modified?

If the PVCR holes have been drilled out by a previous owner there will likely be tuning problems ahead. What can have happened is that a previous tuner will have tuned the engine using the main jets to achieve the ideal full power mixture only then to find that the two main jets are insufficient to run the engine correctly on part throttle. The PVCR holes have then been enlarged to compensate.

The categorising system does rely on the metering blocks being standard and unmodified. Therefore, always look very closely at the PVCR holes in any metering block to see if they look like they've have had a drill bit run through them. Standard original metering block holes' edges are dead sharp, and the drilling internal surface dead smooth. Use a magnifying class to check this detail. Holes that have been over-drilled are never as neat as Holley's originals.

If PVCR holes have been drilled larger by a previous owner, the only way of accurately identifying the metering block is by the numbers stamped on it. The staff at Holley Technical Services might be able to help, but if you can't positively identify the metering block the only thing that can be done is to try it out on the engine.

Also check to make sure that there are no extra holes in the walls of the main jet wells facing the fuel bowl. If your metering block has been modified for use with alcohol fuel, it could have had this alteration carried out. No metering block drilled for alcohol will ever run properly on a gasoline (petrol) fuelled engine.

List Number stamped on the choke tower of the carburetor body. In fact, most metering blocks do not have the body or carburetor List Number stamped on them at all. If the numbers on the metering block, or blocks, are

The two 'power valve channel restriction' (PVCR) holes are arrowed at 'A' and 'A.'

SPEEDPRO SERIES

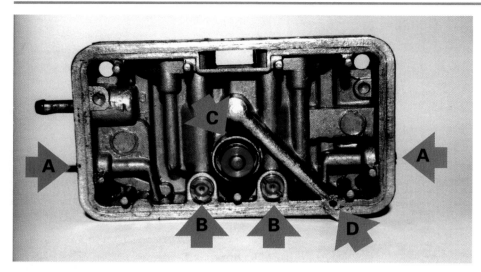

This is the side of the primary metering block that faces the fuel bowl. Idle adjustment screws are at 'A' and 'A.' The main jets are at 'B' and 'B.' The power valve inlet is at 'C.' The accelerator pump entry hole to the metering block is at 'D.'

different from the carburetor's List Number, or different from the numbers Holley Technical Services give you for the List Number of carburetor, then the metering block, or blocks, have been changed at some stage after original manufacture. The question now becomes 'if the metering blocks are not the right ones, what carburetor or carburetors are the metering blocks from?'

The chances are that you will not be able to identify a metering block unless you happen to know from experience that the numbers stamped on it relate to a particular List Number carburetor. There is, unfortunately, no Holley published list to cross-reference metering blocks with carburetors. There is, however, a quick method I can give you to indicate what CFM rating carburetor any particular power valve-equipped metering block has, most likely, come from, or will possibly work on. Essentially the identification process revolves around the fact that the 'power valve channel restriction holes' (PVCR) increase in size as the CFM rating of the carburetor gets larger. Low number CFM rated Holley carburetors have smaller 'power valve channel restriction' holes than larger CFM rated carburetors.

Provided the PVCR holes have not been drilled out at some stage after original manufacture, the hole size is a good guide (but only a guide) as to what CFM rating range of carburetors the metering block is suitable for. That said, the sizes of the holes does vary within the same CFM rated Holley carburetors and some smaller four barrel carburetors have larger PVCR hole sizes than larger CFM rated carburetors, so there's nothing too exact here. The general trend is that the holes go within a range of 0.038 to 0.072in from small to large on 2300, 4150 and 4160 carburetors. The basic PVCR hole sizes range for approximate CFM categorisation is as follows:

2300 Series carburetors
List 7448 - 350-CFM two barrel metering block has 0.059inch diameter PVCR holes

List 4412 - 500-CFM two barrel metering block has 0.062inch diameter PVCR holes

4150 and 4160 Series carburetors
List 8007 - 390-CFM four barrel metering block has 0.038inch diameter PVCR holes
List 4548 - 450-CFM four barrel metering block has 0.038inch diameter PVCR holes
List 1850 - 600-CFM four barrel metering block has 0.049inch diameter PVCR holes
List 4777 - 650-CFM four barrel metering block has 0.041inch diameter PVCR holes
List 4778 - 700-CFM four barrel metering block has 0.053inch diameter PVCR holes
List 4118 - 725-CFM four barrel metering block has 0.070inch diameter PVCR holes
List 4779 - 750-CFM four barrel metering block has 0.056inch diameter PVCR holes
List 7010 - 780-CFM four barrel metering block has 0.062inch diameter PVCR holes
List 4780 - 800-CFM four barrel metering block has 0.069inch diameter PVCR holes
List 9381 - 830-CFM four barrel metering block has 0.070inch diameter PVCR holes
List 4781 - 850-CFM four barrel metering block has 0.067inch diameter PVCR holes
List 8049 - 950-CFM four barrel metering block has 0.069inch diameter PVCR holes
List 8051 - 1000-CFM four barrel metering block has 0.072inch diameter PVCR holes

The foregoing PVCR (power valve channel restriction) diameter hole sizes are as found on primary metering blocks. On four barrel carburetors

BASIC CARBURETOR IDENTIFICATION

the PVCR hole sizes in the secondary metering blocks will be of similar size to the primary ones for CFM rating purposes. In fact, in many instances the primary and secondary metering block will be identical, and will even carry the same part numbers.

Any metering block that has a power valve fitted, PVCR hole sizes smaller than 0.056in in diameter, accelerator pump circuitry and idle adjustment screws is a primary metering block. The reason for this is that only 750-CFM and larger Holley carburetors were fitted with secondary metering blocks which had idle adjustment screws and operative idle circuitry, operative power valves and, on mechanical carburetors, accelerator pump drillings. Using the PVCR holes to categorize metering blocks is approximate only because it doesn't take into account things like the idle and progression phase characteristics (drillings) of the metering blocks.

Is it a primary or secondary metering block?

1. Any metering block that is not machined to take a power valve is a secondary metering block.
2. Any metering block that does not have accelerator pump drillings is a secondary metering block.
3. Any metering block that does not have idle adjustment screws is a secondary metering block.

The foregoing narrows down some of the possibilities for the quick identification of some metering blocks, but it's not as simple as that. Some Holley four barrel carburetors have metering blocks fitted to them which are used by Holley as primary and secondary metering blocks. This means that the primary and secondary barrels have operative power valves, curb (kerb) idle adjustment screws and circuitry and operative accelerator pump passageways on mechanical carburetors. If you are unable to identify a metering block by the numbers stamped on it, categorise it by its PVCR hole diameter sizes (assuming that they are standard and have not been modified after original production).

Most mechanical secondary or 'double pumper' four barrel

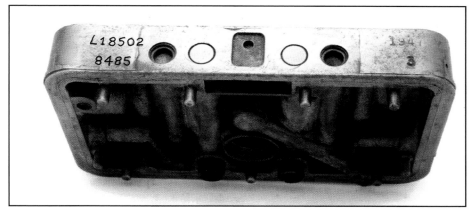

This metering block has three groups of numbers stamped on the top edge. The 'L18502' group is the List Number with '1850' and '8485' being the significant numbers. This is where things get confusing as both of these numbers are applicable. In this instance '1850' is the List Number and '8485' is Holley's individual part identification number. Both sets of numbers identify the metering block.

carburetors do not have secondary metering blocks with curb idle adjustment screws fitted, although **all** 4150 **and** 4160 four barrel carburetors do have curb idle circuitry in them and all will have accelerator pump drillings. A metering block without curb idle adjustment screws is a mechanical secondary one.

Some vacuum secondary 4150

All 2300 two barrel carburetors and all 4150 and 4160 four barrel carburetors have primary metering blocks that look like this from the side that faces the carburetor body. The arrow indicates the power valve.

SPEEDPRO SERIES

four barrel carburetors will have an operative power valve system fitted to them, but the metering blocks for these carburetors will not have accelerator pump circuitry or curb idle adjustment screws and circuitry drillings. A metering block without accelerator pump drillings is definitely a vacuum secondary metering block.

After identifying what type of carburetor a metering block is from, use the size of the PVCR hole sizes to gain an approximation of the CFM rating of the carburetor from which it originally came.

Metering blocks do differ in other areas apart from the diameter of the PVCR holes. There are various other holes which have been drilled to certain sizes and positioned in certain places to give particular fuelling characteristics. Metering blocks might all look the same at a glance, but they most definitely are not. The idle and off idle characteristics of a particular metering block might well be completely unsuitable and untuneable for a particular application. In some cases a metering block which appears to be correct in terms of PVCR holes and other identifying features will just have to be given up on because it will not work well for any other than its originally intended application.

If a metering block is the correct one for the carburetor, but the engine is of smaller capacity than that for which the carburetor was intended, the engine may not perform as well as expected (especially at the top end). This is because the power valve restriction holes are too large for the engine. This is not Holley's fault. This situation occurs occasionally because Holley calibrates the carburetors for an engine of a particular capacity and you, the end user, are required to select the right sized carburetor for the engine capacity you're using.

The usual problem with using a carburetor that is too large for the engine capacity is that the metering block supplies too much fuel when the power valve circuit comes into operation and floods the engine. Reducing the main jet size is no solution because the engine will not run properly on the mains when the power valve is not working. The possible solution to this problem is to use a secondhand (inexpensive) metering block with smaller PVCR holes (0.038 - 0.044in), such as one from a small four barrel Holley. If the idle circuitry is suitable (quite likely) the top end fuelling will be improved (reduced) with the same main jets in place. Failing this, the PVCR hole sizes can be reduced using an aftermarket kit, though this will entail having the metering block sent away to be machined to allow the insertion of smaller pre-drilled replacement jets into the power valve recess.

Note that metering blocks with over-drilled PVCR holes, or extra holes in the main jet wells, will usually prove to be impossible to tune when the carburetor is fitted to an engine of the correct capacity and state of tune.

THROTTLE BODY & SHAFT ASSEMBLIES

Some four barrel carburetors have the List Number stamped on the throttle body and shaft assembly; others have four completely different numbers stamped on them. Note that the underside of the throttle body and shaft assembly may have to be (carefully) wire brushed to enable the numbers to be read.

Fortunately, throttle body and shaft assemblies do not tend to get swapped around, but be aware that it can have happened. Holley can tell you if the throttle body and shaft assembly is correct for the List Number carburetor you have.

Some throttle body and shaft assemblies do not have any numbers stamped on them at all and, in these cases, the diameters of the butterflies have to be measured to ascertain what carburetors they can be used with.

Butterfly diameters in relation to CFM
2300 Series
350-CFM two barrel: $1^{1}/_{2}$in
500-CFM two barrel: $1^{11}/_{16}$in

4150 and 4160 Series (primary & secondary bores)
390-CFM four barrel: $1^{7}/_{16}$in
450-CFM four barrel: $1^{1}/_{2}$in
600-CFM four barrel: $1^{9}/_{16}$in
650-CFM four barrel: $1^{11}/_{16}$in
700-CFM four barrel: $1^{11}/_{16}$in
750-CFM four barrel: $1^{11}/_{16}$in
780-CFM four barrel: $1^{11}/_{16}$in
800-CFM four barrel: $1^{11}/_{16}$in
830-CFM four barrel: $1^{11}/_{16}$in
850-CFM four barrel: $1^{3}/_{4}$in
950-CFM four barrel: $1^{3}/_{4}$in
1000-CFM four barrel: $1^{3}/_{4}$in

Note that the 4150 and 4160 Series 650 through 830-CFM carburetors all use the same size butterfly, the CFM rating being determined by the venturi size in the body of the carburetor. This situation also applies to 4150 Series 850 through 1000-CFM carburetors.

CFM ratings in relation to venturi sizes
350-CFM two barrel: $1^{3}/_{16}$in primary
500-CFM two barrel: $1^{3}/_{8}$in primary
390-CFM four barrel: $1^{1}/_{16}$in diameter primary & secondary
450-CFM four barrel: $1^{3}/_{32}$in primary & secondary
600-CFM four barrel: $1^{1}/_{4}$in primary & $1^{5}/_{16}$in secondary
650-CFM four barrel: $1^{1}/_{4}$in primary & $1^{5}/_{16}$in secondary

BASIC CARBURETOR IDENTIFICATION

700-CFM four barrel: $1^{5}/_{16}$in primary & $1^{3}/_{8}$in secondary
750-CFM four barrel: $1^{3}/_{8}$in primary & secondary
800-CFM four barrel: $1^{3}/_{8}$in primary & 1 7/16in secondary
850-CFM four barrel: $1^{9}/_{16}$in primary & secondary
950-CFM four barrel: $1^{3}/_{8}$in primary & secondary
1000-CFM four barrel: $1^{9}/_{16}$in primary & secondary

The venturi size and butterfly size is the key to identification until we get to the more recent 950 and 1000-CFM carburetors for which the main discharge booster design was altered to allow for improved airflow through the venturis.

The two barrel 2300 carburetors are relatively easy to identify via their butterfly diameters, venturi sizes and throttle body and shaft assembly.

The 350 and 500-CFM 'standalone' carburetors have an operative power valve in the metering block, drillings in the body of the carburetor (for vacuum to operate the power valve), curb idle drillings in the throttle body and shaft assembly, a manual or automatic choke mechanism and an accelerator pump mechanism. The non 'standalone' two barrel carburetors have no accelerator pumps, no metering block (they have a metering plate instead), no accelerator pump mechanism and diaphragm throttle spindle operation; all of which makes them easy to identify.

The four barrel carburetor bodies and throttle body and shaft assemblies are a bit more difficult to identify but, nevertheless, are quite identifiable by whether they have adjustable secondary curb idle or four corner idle circuitry, non-adjustable secondary idle circuitry (very small hole directly under each secondary barrel progression slot), operative secondary power valves and secondary accelerator pump systems making them either mechanical or vacuum in operation. The mechanical linkage to the two rear barrels is also a giveaway here. The mechanical secondary or 'double pumper' carburetor has a full mechanical linkage and two accelerator pump mechanisms, while the vacuum operated carburetor doesn't have a secondary mechanical link or a second accelerator pump mechanism.

Note that all four barrel carburetor throttle body and shaft assemblies are drilled as if the power valve will be operative. The carburetor body appropriately drilled is what governs whether or not the carburetor has an operative power valve.

Chapter 3

Choosing the right size carburetor

It's confusing that all sorts of carburetor sizes are seen on the same type and capacity of engine. A potential problem here is that people see a carburetor, or carburetors, with a big CFM rating being used on the same engine they have and can see no reason why they shouldn't use the same carburetor. This is the 'bigger is better' syndrome coming into the equation.

Optimum carburetor size depends on the cubic capacity of the engine and its efficiency, the maximum rpm the engine will turn and the individual application.

Very efficient engines turning high rpm will definitely be able to use a large carburetor to develop good top end power (6500rpm and above). This may be at the expense of good low end power (4000rpm and lower). These factors are all taken into account when an engine's true specifications are put into the Holley calculation.

Most poor performance problems (where they're carburetor related) stem from carburetors that are too large. If the carburetor is too small, the lack of carburetor capacity will only become apparent at the top end of the particular engine's rev range. The power will flatten off rather abruptly as the carburetor just cannot flow sufficient air to meet the engine's needs. However, up to this rpm barrier, the engine will perform perfectly satisfactorily.

A simple calculation using several factors is used to work out the CFM requirements for the carburetor for a particular engine.

There are 1728 cubic inches in one cubic foot (for those working in cubic centimeters 1000cc is approximately 61 cubic inches).

A factor of the Holley calculation is the volumetric efficiency rating of the engine (knowing exactly how much air it can flow). The volumetric efficiency of an engine depends on how well the engine has been designed and/or modified. This is where it all gets a bit difficult because the engine may not be as efficient as you think, meaning that the over optimistic figures you put into the calculation are not correct and the resulting answer is then wrong too. An over estimate of volumetric efficiency usually leads to a larger carburetor than is desirable being fitted. Conversely, an under estimate could lead to an undersized carburetor being fitted.

Your particular application is also very important because it is essential to have the power where you need it in the rev band (i.e. - mid-range or top end only). If the result of the calculation, expressed in CFM, is a figure that is in the middle between two carburetor sizes, go for the larger size if top end power is more important and go for the smaller size if mid-range power is more important.

Using the Holley calculation will prevent a gross mismatch provided the volumetric efficiency rating used is

CHOOSING THE RIGHT SIZE CARBURETOR

realistic and the maximum revs which will be used on a regular basis are not over estimated. Too many people appear to use a 100% volumetric efficiency rating in their calculation and this mistake in one variable leads to all sorts of complications.

The object of the exercise is to use a carburetor with exactly the right amount of CFM airflow, no more and no less. Badly mismatched carburetor/engine combinations can be the cause of many tuning problems. When the carburetor is too large, many engines will have a large 'flat spot' in the acceleration phase. The engine accelerates reasonably well, but only really gets going higher up the rpm range. In such circumstances, when a smaller carburetor is tried, the engine livens up considerably and is very often only slightly inferior at the top end of the rpm range to how it was with the larger carburetor.

The following Holley calculation examples give a fair range of engine sizes from the smallest to the largest V8, with commonly used rpm ranges and efficiency ratings that are reasonable. The variation of carburetor sizes potentially applicable to a particular engine size is also clearly demonstrated. Obviously, you can only get an accurate answer from the calculation if the figures used for the variables are reasonably accurate.

CFM CALCULATION

1 - Multiply half the cubic inches of the engine's total capacity by the maximum rpm to be used.
2 - Divide this answer by 1728.
3 - Multiply this answer by the efficiency rating of the engine (e.g. - 75% efficiency meaning that 0.75 is the multiplication factor)
4 - The answer is the cubic feet of air per minute (CFM) that the engine can flow, and the choice of carburetor is based on this amount of airflow.

3500cc engine at 5000rpm - 75% efficiency = 233-CFM use (350-CFM)
3500cc engine at 6500rpm - 85% efficiency = 349-CFM use (350 to 500-CFM)
3500cc engine at 7000rpm - 95% efficiency = 413-CFM use (450 or 500-CFM)
5000cc engine at 4700rpm - 75% efficiency = 349-CFM use (350 or 500-CFM)
5000cc engine at 6500rpm - 85% efficiency = 481-CFM use (500 or 600-CFM)
5000cc engine at 7200rpm - 95% efficiency = 597-CFM use (600-CFM)
5000cc engine at 7800rpm - 95% efficiency = 647-CFM use (650-CFM)
5700cc engine at 5000rpm - 75% efficiency = 379-CFM use (390, 450 or 500-CFM)
5700cc engine at 6500rpm - 85% efficiency = 559-CFM use (600-CFM)
5700cc engine at 7200rpm - 95% efficiency = 692-CFM use (700-CFM)
7400cc engine at 4500rpm - 75% efficiency = 443-CFM use (450 or 500-CFM)
7400cc engine at 6500rpm - 85% efficiency = 725-CFM use (725 or 750-CFM)
7400cc engine at 7000rpm - 95% efficiency = 873-CFM use (850-CFM)

This panel shows a range of common V8 engine capacities (in cubic inches), maximum revs (in rpm) and correct efficiency ratings (expressed in percentages) with the resultant peak flow figures (the carburetor size to use is in brackets).

If you're wondering why just half the engine's capacity is used in the calculation, it's because we're dealing with four stroke (otto cycle) engines, and only every second revolution of the engine per cylinder is to do with drawing air into the engine. The next part of the calculation divides the answer by 1728 because 1728 is the number of cubic inches in a cubic foot. The final part of the basic calculation is multiplying this answer by the volumetric efficiency rating of the engine and this is where the gross errors are likely to creep in. The efficiency rating of an engine should be realistic, not over rated.

If the foregoing examples don't apply to your engine and application, simply substitute your actual variables and a realistic volumetric efficiency rating percentage to get a CFM to use in carburetor choice. Do bear in mind the application because, if mid-range torque is required, then a smaller CFM carburetor from your choice range will almost always prove to be more suitable. The possible variation in CFM rating can be quite significantly influenced by the volumetric efficiency figure you use and there is no point in calculating the carburetor sizing at 95% efficiency if your engine is stock and likely to be 75% efficient at most.

With the two barrel carburetors, it's better to have a 350-CFM one working up to maximum efficiency than it is to have a 500-CFM one working at up to 60% efficiency. With a four barrel mechanical carburetor, it is better to have all four barrels working up to maximum efficiency than all four barrels of a mechanical carburetor working at 60% efficiency.

Although the Holley formula is usually extremely reliable, a further way of refining carburetor size for optimum performance is to use the Holley CFM calculation and then obtain two different sized carburetors and try them both. This is not an outlandish idea, if you have a range of Holley parts and are able to make up two different sized carburetors. Testing an engine with two different sized carburetors can often be quite revealing: the real world application dynamics deciding which carburetor is ultimately going to be better to use.

The volumetric efficiency rating of a naturally aspirated engine is related

SPEEDPRO SERIES

to how full the cylinders are after the induction cycle. Here are some examples you can use to decide the efficiency of your engine:

1 - Average stock engine: 70 to 75% efficient.
2 - Mildly modified engine with reasonably well ported stock cylinder heads, a 280-285 degree duration camshaft, tubular exhaust extractor system and the other usual mods and equipment seen on engines like this: 80% to 85% efficient.
   3 - Fully modified engine with aftermarket cylinder heads, a 290-300 degree duration camshaft and all of the best bits: 85% to 95% efficient. Using the 100% figure is clearly not right for at least 95% of Holley carburetor equipped engine applications.

Take it that, for calculation purposes, the average stock engine is going to be 75% efficient, the average mildly modified engine is going to be 85% efficient and that the average racing engine is going to be 95% efficient for carburetor size calculation purposes. Beyond this, testing each applicable carburetor out on the particular engine is the only way to narrow down what really is the absolutely right carburetor for a particular engine and application.

It's quite possible to fit a 500-CFM two barrel carburetor on to a 7 litre V8 engine. Such an engine will run perfectly up to a certain rpm. In fact the engine will run unbelievably well up to the point that the carburetor reaches its maximum air flow capability. This will happen at about 5000rpm or so on a stock engine and slightly higher if the engine is fitted with a good proprietary small runner 360 degree inlet manifold. What happens is that the engine gets to a particular rpm and the 'surge' simply stops. In fact, on an under-carbureted engine like this, even going downhill will not usually see the rpm increase because the engine is effectively 'governed' by the maximum airflow of the carburetor. The only serious problem with fitting a carburetor that is seriously too small is the fact that the engine could go into 'vacuum' at high rpm and, as a result, the power valve shuts off causing the mixture to go lean (engine damage possible).

Resist the temptation to fit a carburetor with a larger than necessary CFM rating.

www.velocebooks.com/www.veloce.co.uk
All books in print • New books • Special offers • Newsletter

Chapter 4
Basic carburetor parts

REPAIR KITS

Genuine Holley repair kits are very comprehensive. Several aftermarket companies also make very good quality repair kits, though it is recommended that only genuine Holley power valves ever be used in Holley carburetors. The reason for this is that many aftermarket replacement power valves do not seem to work as well, or be as consistent in operation, as the genuine Holley components. The stock repair kits come with 6.5Hg power valves and 30cc accelerator pump diaphragms.

The genuine repair kit comes with every washer, every gasket, an instruction sheet and a parts listing.

Warning! Replace every gasket when rebuilding a Holley carburetor.

Fuel leaks are dangerous. Avoid the temptation to re-use old gaskets, even though they may appear to be in perfect condition.

Unpacked contents of a genuine stock Holley carburetor repair kit (sticky, use once metering block to carburetor body/fuel bowl gaskets).

Genuine Holley carburetor repair kit.

SPEEDPRO SERIES

Although the repair kits are very comprehensive, there are, of course, no spare gaskets included. You simply have to have on hand certain spare

Keep on hand a spare fuel bowl gasket for each fuel bowl, they are quite delicate.

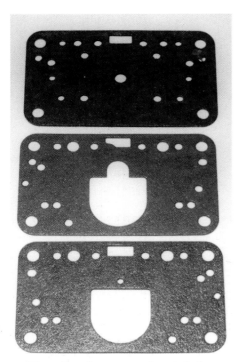

There is quite a range of gaskets for Holleys. All can be identified - even if loose as shown - by matching the shapes to the diagrams in the 'metering block gasket listing' in the Holley Performance Parts Catalogue. This catalogue covers this aspect of Holley carburetors completely and even gives part numbers.

gaskets if you are going to tune your own Holley carburetor. The one thing everyone becomes aware of sooner or later (usually sooner) with these Holley carburetors is the fact that when the float bowls are removed to make a jet change for instance, the fuel bowl gasket is invariably ruined. The later gaskets are all 'sticky' and are 'fit one time' gaskets. They seldom leak, but, of course, if you are in the process of jetting your engine the float bowls may well have to come off and go on again ten times in one day. While you might get two or three successive fittings with a single gasket of this type that's usually about it.

Genuine Holley and aftermarket manufacturers' fuel bowls, metering blocks, metering block to body and metering plate gaskets are available singularly from Holley agents and parts supply shops. Whether the gaskets are genuine Holley or aftermarket ones does not seem to matter as all seem to be of good quality.

The *Holley Performance Parts Catalog's* Numerical Listing gives what gaskets each List Number carburetor needs by way of a part number. Elsewhere in the catalog there are part number listings with diagrams for metering block gaskets, fuel bowl gaskets and metering plate gaskets.

If the carburetor you have is not listed in the Numerical Listing take the body and throttle body and shaft assembly, metering block or blocks and float bowls to a Holley parts dealer. The shop staff can then match gaskets to components: note the part numbers they give you for future reference. There's nothing worse than finding that the gaskets that you have are the wrong ones especially when the carburetor is in pieces at a race venue. Of all the Holley carburetor gaskets, by far the most fragile are the fuel bowl to metering block gaskets.

Warning! Failure to fit a new fuel bowl gasket or metering block gasket each time after removing a fuel bowl constitutes a risk of an engine fire through serious fuel leakage. A perfect fuel seal is possible provided the gasket is fitted correctly. The frequent claim that Holley fuel bowls always leak is not justified, it's almost always installer error that leads to leakage. Carburetor bodies have been known to become

Light blue coloured reusable metering block to carburetor body and fuel bowl to metering block gaskets in their usual packaging.

BASIC CARBURETOR PARTS

Rear view of this vacuum secondary four barrel carburetor body shows that it cannot take a metering block fitted with a power valve as there's no appropriate recess.

Rear view of this vacuum secondary carburetor body shows that it can take a metering block fitted with a power valve as it has a recess. Check that the carburetor body has been drilled so that the power valve will operate. This one has not been drilled so, while it will take a power valve, the valve will not actually work. Arrow 'A' indicates where to look for a drilled hole.

This is a view of the primary side of a four barrel carburetor. ALL of these 2300, 4150 and 4160 series Holley carburetors have provision for a front mounted power valve and the hole drilled to make the power valve circuit operative. Arrow 'A' indicates where the vacuum transfer hole is drilled.

warped and this can be the cause of constant fuel leakage, but it's not a common problem.

If you are setting up your carburetor and the bowls are coming off regularly, it can get a bit annoying ruining a gasket each time the fuel bowl comes off. In this situation grease can be applied to the surfaces of a new 'one time' gasket (if that's what you're using) which will reduce the 'stickiness' of the gasket allowing it to be removed several times during testing, provided grease is re-applied each time the gasket is refitted. Once testing has finished, a new gasket that has not been greased can be fitted for a perfect seal. The gaskets that were used for testing can be re-greased and kept in an airtight plastic bag for future temporary use. Throw them away if they have any signs of damage.

Holley also makes reusable, 'non-sticky' competition gaskets, which are light blue in colour and come in a specific rebuild kit. These gaskets are durable enough to be reused about 20-30 times or more. Most racers use these gaskets to facilitate jetting changes without having to replace the gaskets.

FINDING THE LIST NUMBER

The List Number is generally required for ordering repair kits and spares. Once the List Number, as stamped on the main body of the carburetor, has been found (you may have to clean the carburetor to read it clearly), the airflow capability (CFM rating) of the carburetor can be found by checking the List Numbers against the *Holley Performance Parts Catalog's* Numerical Listing. Because of the extremely large number of models of carburetor that have been made by Holley over the years many List Numbers will not be found in any partial listing.

The identification process has

SPEEDPRO SERIES

The List Numbers are stamped on the choke towers of the bodies of all of these Holley carburetors in the place shown. This 'List-3259' coding means that this is a 725CFM vacuum secondary four barrel carburetor.

Mechanical choke linkage is simple and effective, and virtually never ever goes wrong. It also generally never needs to be used. Manual chokes are either all metal like this one or injection moulded in black plastic.

to be taken further on four barrel carburetors because of the various differences on their secondary side. A good example of this is whether or not the carburetor body has provision for a secondary power valve. Any carburetor body that can take a power valve will have the cast in recess in the back of the body to take the head of the power valve. However, not all carburetor bodies are drilled by the factory for secondary power valve operation, meaning that while they can take a power valve they were not factory designed to work with one. A visual inspection when the carburetor is in pieces will help identify this situation.

CHOKE

If you are buying a new carburetor, it's probably better not to choose a Performance Holley carburetor fitted with an automatic choke. While these choke mechanisms work well when new, they always seem to cause operating problems as they get older (and they are often not that old before problems arise!).

The usual problems are that they don't switch on or don't switch off, or take far too long to switch off. The solution to this problem (if the use of a choke is required) is not to persevere with the automatic choke but to fit a manual choke conversion. Having a manual choke is not normally a major disadvantage. **Note.** Automatic chokes are not covered in this book.

FUEL BOWLS

There are two basic types of fuel bowl for these Holley carburetors. They

Primary 'centre pivot'-type fuel bowl has the fuel inlet on the left 'A' and the sight plug at 'B'. The accelerator pump is on the underside of the fuel bowl and on the right. The shape of the centre pivot fuel bowl is quite distinctive.

BASIC CARBURETOR PARTS

Primary 'side hung'-type fuel bowl has the fuel inlet on the right at 'A' and the sight plug at 'B' on the left-hand side of the fuel bowl. The accelerator pump is on the underside of the fuel bowl and on the the right 'C'. The shape of the side hung fuel bowl is quite distinctive.

are 'side hung' and 'centre pivot' types. The description of the fuel bowls as being either 'side hung' or 'centre pivot' refer to the direction that the float works in the fuel bowl. The floats are positioned in the fuel bowls 90 degrees apart in each case. The differences between the fuel bowl designs is all about maintaining the fuel level in the fuel bowls under different circumstances. The arc of travel of the float action is across the car with 'side hung' fuel bowls and lengthways along the car's axis for 'centre pivot' fuel bowls.

Note that some needles and seats are non-adjustable on some 'side hung' fuel bowls: the information sheet as found in the Holley parts kit explains exactly how to set these fuel bowl's float levels.

Two and four barrel carburetors can be fitted with either type of fuel bowl.

Note that all examples of fuel bowls and their position/orientation as mentioned in this book are based on the orientation of the carburetor when its List Number (stamped on the choke tower) is facing you. If the carburetor has no choke tower the main throttle linkage connection is on the right-hand side of the carburetor and nearest to you when the body is in the same orientation.

'Side hung' fuel bowls

'Side hung' fuel bowls are the most common with the only difference between two barrel or four barrel carburetors being the fact that the fuel bowl for a two barrel carburetor does not have provision for a fuel feed transfer tube going back to a secondary while a four barrel's does. It's not possible mix up 'side hung' fuel bowls for 2300 and the primary side of 4150 or 4160 carburetors for this reason. The secondary 'side hung' fuel bowl for a four barrel carburetor is opposite handed to the primary fuel bowl.

A dummy assembly of the fuel bowls onto the carburetor body with

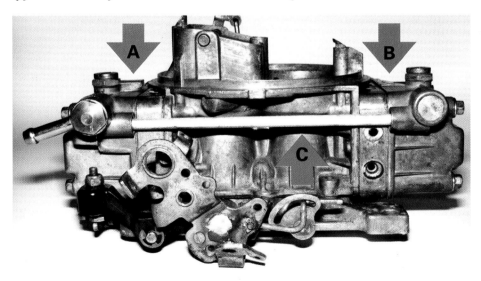

The primary 'side hung'-type fuel bowl on the left 'A' feeds fuel back to the secondary 'side hung' fuel bowl 'B' via a fuel feed transfer tube 'C'. It is not possible to confuse the positioning of 'side hung' fuel bowls on any carburetor. Two barrel fuel bowls have no operative drilling for a fuel feed transfer tube and four barrel 'side hung' fuel bowls are handed as shown in this photo.

SPEEDPRO SERIES

General configuration of primary and secondary handing of 'side hung'-type fuel bowl inlets on a vacuum secondary four barrel carburetor. The primary fuel bowl is on the left-hand side and the secondary is on the right-hand side. The float level adjusting screw mechanism of the primary fuel bowl is bottom left at 'A,' while the secondary float level adjusting screw mechanism is bottom right at 'B.' There is no possibility of confusion.

The primary 'side hung' fuel bowl of a four barrel carburetor is quite easy to distinguish from a secondary 'side hung' fuel bowl of a four barrel carburetor. The primary fuel bowl has a right-hand fuel inlet union, a filter system and rear facing fuel feed take off fitment for a transfer tube. The fuel level 'sight plug' is on the left-hand side. The float level adjustment is positioned on the right-hand side of the fuel bowl if the floats are externally adjustable. The accelerator pump is on the right-hand side and on the underside of the fuel bowl.

The secondary 'side hung' fuel bowl has a right-hand transfer tube fitment and a slightly different casting boss which is plugged (as opposed to being threaded to take a union fitting). There is no provision for an accelerator pump and the sight plug is on the left-hand side of the fuel bowl. The float level adjustment is positioned on the right-hand side of the fuel bowl if the floats are externally adjustable.

An alternative method of identifying a primary 'side hung' fuel bowl is as follows: If you hold it and look at it with the Holley lettering on the front of it facing you, the fuel inlet boss is cast in and is positioned on the right-hand side of the fuel bowl: the sight plug will be on the left-hand side of the fuel bowl. The float level adjustment screw mechanism will also be on the right-hand side of the fuel bowl. The accelerator pump will be on the right-hand side on the bottom side of the fuel bowl. The back of the fuel inlet boss on the right-hand side of the fuel bowl will be machined to take the end of a transfer tube.

On the other hand, a secondary 'side hung' fuel bowl with the Holley lettering facing you will have the fuel inlet of the transfer tube on the left-hand side of the fuel bowl, the sight plug on the right-hand side of the fuel

the body and throttle body and shaft assembly in the correct orientation will allow the correct identification of all fuel bowls.

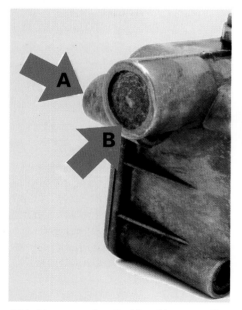

This is a secondary fuel bowl because the inlet at A is on the lefthand side of the fuel bowl and there is a plug at B meaning that there is no provision for a fuel inlet fitting.

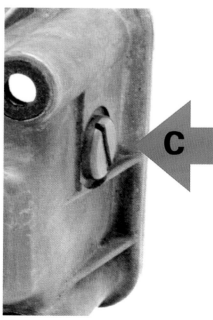

On a secondary 'side hung'-type fuel bowl the sight plug is on the right-hand side at 'C.'

BASIC CARBURETOR PARTS

Usual 'side hung'-type primary fuel bowl with the fuel union boss on the right-hand side at 'A' and the sight plug on the left-hand side 'B.' The accelerator pump is on the bottom right-hand side of the fuel bowl at 'C.' The float level adjustment screw mechanism is at 'D.'

'Centre pivot'-type fuel bowls for four barrel Holley carbs have left-hand fuel inlets (arrowed at 'A' and 'A') and a left-hand positioned sight plug (arrowed at 'B' and 'B'). It not possible to mix them up.

except a drilling to take a transfer tube, although it will have the cast boss to take a fuel union. If you pick up a 'side hung' fuel bowl and look at it with the Holley lettering on the front of it facing you, the fuel inlet boss is cast in and positioned on the left-hand side of the fuel bowl, and the sight plug will be on the right-hand side of the fuel bowl: there is no provision for an accelerator pump.

'Side hung' fuel bowls for four barrel carburetors can be paired up easy enough with the transfer tube being used to connect them. The transfer tube is always on the right-hand side of the carburetor when you're facing the List Number (stamped on the choke tower). The choke is on the opposite side of the carburetor as is the vacuum diaphragm housing (if it's a vacuum secondary carburetor). The sight plugs will be on the left-hand side of the carburetor in this situation. A dummy assembly can be used to ascertain the exact fitting of fuel bowls if there is any doubt.

There are some variations on secondary 'side hung' fuel bowls but they are relatively rare. For instance, some 'side hung' fuel bowl equipped four barrel 'double pumper' carburetors have a primary fuel bowl on the secondary side of the carburetor. This means two separate fuel lines going to the carburetor from the fuel pump; one fuel pipe going to the right-hand side of the carburetor to feed the primary fuel bowl and the other fuel pipe going to the left-hand side of the carburetor to feed the secondary fuel bowl. The sight plug holes are also on different sides of the carburetor. The giveaway here is whether or not the end of a transfer tube can be fitted to the fuel bowl (these ones can't). Fuel bowls like this are similar to those from a 350 or 500-CFM two barrel carburetor and they

bowl and the float level adjustment screw mechanism on the left-hand side of the fuel bowl.

A 'side hung' fuel bowl which is for use on the secondary side of a carburetor has no fuel inlet system both have accelerator pumps which is a further means of identification. A four barrel carburetor, for example, which has this separate fuel inlet arrangement to each fuel bowl is the 750-CFM double pumper (List Number 6109). These were racing carburetors and relatively few were made.

'Centre pivot' fuel bowls

All the 'centre pivot' fuel bowls are virtually the same when originally die cast by Holley, no matter whether they're going to be for a two barrel or a four barrel carburetor primary or secondary. However, after the factory has machined these fuel bowls they'll either be suitable solely for fitting to a two barrel carburetor or primary fitting (same thing) or secondary fitting on a four barrel carburetor. Primary and secondary 'centre pivot' fuel bowls have a left-hand fuel inlet and a left-hand fuel level sight plug when fitted to four barrel carburetors, making them instantly distinguishable from each other (if they're mixed, these features will be on the opposite side of the carburetor). Only one side of the fuel bowl will be machined to take a fuel union and, in each case, the machined side of the bowl will be opposite.

There is an exception to this overall uniformity: some four barrel carburetors made for Ford some years back were fitted with centre pivot fuel bowls which had a single fuel inlet pipe going to the left-hand side of the primary fuel bowl, and a transfer fuel pipe on the right-hand side (feeding to the right-hand side of the rear fuel bowl). The sight plug is positioned on the left-hand side of the rear fuel bowl. In this case both sides of the primary fuel bowl were machined to take fuel pipe unions (cross-feed drilling).

The 'centre pivot' float bowl was developed for racing to cope with cornering G-forces. What happened with the 'side hung' float bowls was that when a car leaned into a curve, this effect, coupled with G-forces, would cause the fuel in the fuel bowl to move to one side. Then, depending on which way the car was turning, the float would either drop or rise. If the float dropped, it would open the needle valve an excessive amount and let more fuel into the fuel bowl. When the car levelled off after negotiating the turn, so would the fuel, but at a higher level than normal, causing a rich mixture until the excess fuel was burnt off. Conversely, turning in the opposite direction would cause the float to rise, reducing fuel supply and leaving a low fuel level for a short while when the cornering forces subsided. In both circumstance the engine would end up misfiring, which is not the quickest way around a race track. Holley introduced the 'centre pivot' float bowl to overcome this fuel surge problem, which serves to emphasise the importance of maintaining the correct float level (and therefore fuel level) and using the correct fuel bowl design for the application.

'Centre pivot' fuel bowls on four barrel Holley carburetors normally have a twin feed fuel inlet piping system.

Not all secondary 'centre pivot' fuel bowls have an operative accelerator pump. Any 'centre pivot' fuel bowl that does not have provision for an operative accelerator pump mechanism is definitely a secondary fuel bowl from a vacuum secondary carburetor.

Caution! It is extremely easy to strip the threads in the body of the carburetor when fitting the fuel bowl securing screws. The golden rule is to offer the fuel bowl up to the carburetor, and then wind in all four screws by hand, but don't tighten any of them until they are all more or less fully wound in. If the two top screws, or the bottom ones for that matter, are fully tightened before the remaining two are fitted, a slight misalignment are possible and could result in a ruined thread. With the fuel bowl firmly fixed in position by two screws, the two remaining screws will often start in the threads, and wind in, but the loading may be to one side of the thread, and the screw, being steel, can strip the thread in the body.

To prevent this problem, wind all four screws in by hand until they are virtually seated before tightening with a screwdriver. If any resistance is felt, take the fuel bowl off and identify the cause.

FLOATS

There are three types of float material used by Holley. Floats can be made of brass, or Nytrophyl or Duracon injection moulded plastic. The 'side hung' fuel bowl has no clearance problems irrespective of what type of float is fitted. There are two types of brass float for 'side hung' fuel bowls.

There are two lengths of 'centre pivot' float and, in one instance, the correct one definitely has to be fitted. If a 'centre pivot' float bowl is being fitted to a carburetor which has a secondary metering plate, the 'short float' must be used because it is designed to clear the metering plate. The 'short float' can, however, be used in **all** 'centre pivot' fuel bowls, while the 'long float' is for all 'centre pivot' fuel bowls which are used in conjunction with metering blocks. The only reason for using a 'long float' (if you can) is the easier shutting off and opening action of the needle and seat that they afford. Though, that said, the 'short floats' seldom, if ever, seem to be the cause of any fuel level problems.

Note that some Holley 'side hung' non-performance application

BASIC CARBURETOR PARTS

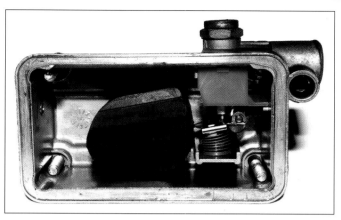

Nytrophyl float as fitted to this 'side hung'-type fuel bowl.

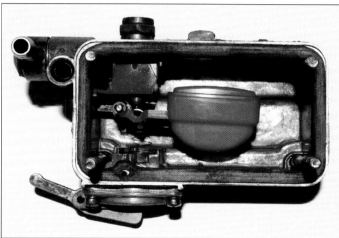

Duracon float as fitted to this 'side hung'-type fuel bowl.

Long pivot brass float on the left and short pivot Nytrophyl float on the right. The 'short float' on the right is for fitting into centre pivot fuel bowls which are used in conjunction with metering plates. The 'long floats' will not fit in the same situation.

POWER VALVES

The power valve is a valve that comes into action when there is a sudden demand for engine power (richer mixture required) such as when pushing the accelerator pedal to the floor during full acceleration. All primary, and some secondary, metering blocks have a power valve. The power valve allows extra metered fuel to flow from the fuel bowl into the main jet well in the metering block when it opens, and then into the main discharge nozzle.

The power valve uses engine vacuum to keep the valve closed (no fuel can flow) and, conversely, a lack of vacuum to open the valve and let fuel flow. When the engine is switched off, the power valve is open but because the engine is not running no fuel flows. Consider the power valve to be a vacuum operated switch mechanism which will open in moments of sudden power demand (low inlet manifold vacuum) and remain firmly shut at all other times (high inlet manifold vacuum).

Each power valve has a rated opening point in inches of vacuum (Hg) stamped on it. Power valves are made in Hg ratings from 2.5Hg to 10.5Hg in various increments (0.5, 1.0 and 1.5Hg). So, if a carburetor has a power valve fitted to it with 6.5 stamped on it (either on the face of the diaphragm cover or the edge of the die-casting), it is rated to open when the engine vacuum drops below this amount of inlet manifold vacuum.

The higher the numerical numbering of the power valve (10.5, for example), the sooner it will open meaning that extra fuel will start to flow into the engine quite early in the acceleration process (or from a relatively small amount of accelerator pedal depression). The lower the numerical numbering of the power

fuel bowls do not have externally adjustable floats. There's nothing wrong with these fuel bowls, provided the float levels are set correctly. However, because the adjustments are not external, accurate setting is more difficult so these bowls are not really recommended for use on a high performance engine.

SPEEDPRO SERIES

The three common types of power valve. Later style 4 hole power valve on the left, older style 4 hole type in the centre and current 'window type' on the right. Only 'window' types are made at the time of writing.

'50' stamped (arrowed) on the side of this power valve means it's a 5.0Hg one.

valve (2.5, for example), the more throttle opening required before the power valve opens and allows more fuel into the engine. The instant the accelerator pedal is pressed to the floor all power valves will open.

Note. Always start with the recommended standard power valve fitted to the carburetor: it is just amazing how frequently this is the best one to use.

Holley list two 'high fuel flow' power valves with 6.5 and 10.5Hg ratings for use with metering blocks with large power valve restriction holes (part numbers 125-165 and 125-1005). These two power valves will normally only be used on large CFM carburetors fitted to large capacity engines, and are designed to be used with metering blocks which have power valve channel restriction hole sizes of 0.097-0.128inch in diameter. Check the size of the power valve restriction holes using the shank of a 0.098inch diameter size 40 number drill. If the drill shank goes into the holes then you could be limited to these two power valves, which is fine if your engine requires this much fuel. If it doesn't, the use of the other non 'high fuel flow' power valves will limit the amount of fuel supplied and will mean that the main jets will have to be larger than they otherwise would be to achieve the correct high rpm full power air/fuel mixture.

The amount of fuel that flows into the engine once the power valve is open is designed to be controlled by the two holes in the power valve

Numbers '6' and '5' stamped on the pressed steel metal facing of this power valve meaning it's rated at 6.5Hg.

BASIC CARBURETOR PARTS

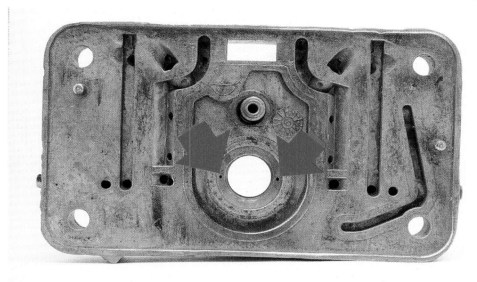

This primary metering block is from a 390-CFM carburetor and has very small power valve channel restriction holes.

This primary metering block is from a 780-CFM carburetor. The power valve channel restriction holes are nearly as large as they get.

particular engine). Holley generally get this absolutely right for almost every application.

While the metering blocks might all look the same, except for the size of the power valve channel restriction holes, there are can be many other subtle differences. This can mean that even if two metering blocks have similar size power valve channel restriction holes, the performance of each metering block on the same engine can be quite different. Make sure that the primary metering block on your carburetor is the right one for the carburetor!

The power valves to use in 2300, 4150 and 4160 Holley carburetors are the single stage ones only. There have been two types of single stage power valve, those with rectangular slots or 'windows' for the fuel to pass through, or those with a series of holes drilled in them. The power valves with holes are the older ones. All current single stage power valves have 'windows'.

To check for freedom of movement in the actual valve action of the power valve, push the spindle inwards (that's the brass cap and spring assembly of the valve that is normally situated in the fuel bowl) while looking at the diaphragm material on the opposite side of the power valve. The diaphragm rubber must not show signs of having cracks on the surface and the valve must return near instantaneously to its normal position. If the rubber has gone hard the valve will not return to its normal closed position quickly: a power valve is definitely not serviceable in this condition and must be replaced.

ACCELERATOR PUMP

The 500-CFM two barrel carburetor, for instance, comes fitted as standard with a 50cc per ten strokes of the pump accelerator pump. While this

mounting recess in the metering block. The fuel has to flow through the power valve as well, of course, but when the standard recommended power valve for the metering block is fitted the flow rate of the power valve will not be a restricting factor. This leads to the situation of having to have the right metering block for the carburetor (suitably sized holes to deliver the extra fuel to the main jet well to provide the right air/fuel mixture ratio for the

SPEEDPRO SERIES

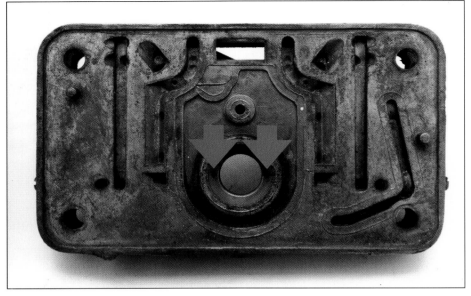

The two power valve channel restriction holes (arrowed) which are specifically sized by Holley during manufacture, control the amount of fuel that flows into the main jet wells.

Later style 4 hole power valve with the flat gasket surface uses triangular central section washer (left). The current 'window' type of power valve with large plain washer (centre). Older style 4 holed type of power valve which has an outside location ridge and uses a gasket which has three tabs (right).

accelerator pump discharge nozzle during acceleration).

Engines accelerate best when they are fed the correct amount of fuel rather than the maximum amount of fuel that the carburetor can supply. Taking the trouble to fine tune the accelerator pump shot size to suit the particular engine results in optimum acceleration. A lean accelerator pump shot, or poorly adjusted pump mechanism results in hesitation. A rich accelerator pump shot results in excessive black smoke from the exhaust pipes and a sluggish accelerating engine. The engine has to get enough fuel, but in the accelerator pump's action phase it does not need more fuel than enough.

Do not fit a 50cc accelerator pump unless it proves to be necessary to use one. Any of these Holley carburetors can be fitted with a 50cc accelerator pump, but the more common 30cc one is usually enough. Stay with a 30cc accelerator pump unless there is a definite requirement for a larger one.

The physical size of the 50cc accelerator pump body and, more specifically, the actuating mechanism arm may mean that the carburetor has to be packed up away from the inlet manifold to give sufficient clearance to the actuating arm. The alternative here is to grind a clearance slot into the inlet manifold, or you can use a combination of the two things to achieve the required clearance. Without sufficient clearance the arm cannot move. Check this feature when fitting a carburetor equipped with a 50cc accelerator pump to any stock type inlet manifold.

In some instances it may prove beneficial to fit a 50cc accelerator pump on the primary side of a four barrel carburetor and a 30cc accelerator pump on the secondary side. There is room for some mixing

might well be excellent for an all out competition engine, this amount of accelerator pump 'shot' is seldom necessary for a road going engine of any size. The more usual 30cc per ten strokes accelerator pump is usually preferable in the interests of fuel economy and providing just the right amount of fuel to accelerate the engine provided the engine's response is ideal (no hesitation caused by a lack of fuel volume being supplied to the

BASIC CARBURETOR PARTS

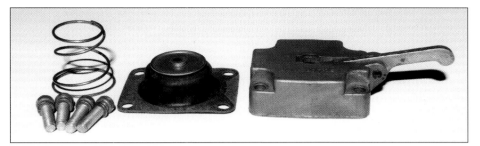
50cc accelerator pump body, diaphragm, return spring and securing screws.

Deeper 50cc accelerator pump diaphragm (above) and shallower 30cc diaphragm (below). Note the distinctive style of the 50cc accelerator pump body.

30cc accelerator pump body is much shallower than 50cc pump body.

30cc accelerator pump diaphragm, body, return spring and four securing screws.

The range of accelerator pump cams are available in a single pack.

and matching here to obtain ideal accelerator pump actions.

ACCELERATOR PUMP CAMS

There are nine different accelerator pump cams and each cam is a different color for easy identification and the cams are also numbered. The cams are made of plastic and are injection moulded. The colors and numbers are in order of fuel delivery, going from the smallest amount (pink) to the largest amount (yellow).

Each cam gives a different fuel delivery rate curve. This adds up to quite a range of fuel delivery actions. The delivery of fuel is not necessarily *pro rata* and, while the Holley literature lists each cam's performance output in ccs (cubic centimetres) per ten full strokes, this is not a true indication of the operating characteristics of the cams. The factory output listing is a guide only and the only way to optimise cam choice is by trial and error, and narrow down the requirements especially when the 'shot' size really is being narrowed down to fine limits.

The other significant part of each accelerator pump cam is that each cam has two location holes moulded into it (a frequently missed point). The cams can all be positioned on the throttle spindle linkage plate in one of two positions. Each position offers a different accelerator pump action. A single screw is used to secure the cam in position. The screw 'self taps' into the plastic the first time the screw is screwed into each particular hole. Make sure that the screw is dead square when it is first screwed into the cam. This facility creates 18 accelerator pump cam actions in total to choose from. The 9 cams can be ordered in a single pack.

With the accelerator pump cam set in the 1 position the total amount

SPEEDPRO SERIES

30cc pump
Pink for the 330
Red for the 240
Green for the 290
White for the 248
Black for the 234
Orange for the 466
Blue for the 427

50cc pump
Brown for the 336
Yellow for the 643.

The primary spindle linkage plate has two holes in it for accelerator pump cam location. Positions are stamped '1' and '2.'

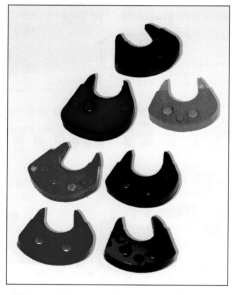

Seven of the possible nine accelerator pump cams. Each of the nine cams has a different profile.

of fuel delivered, and the rate of fuel delivery, will generally be less (leaner) than when that same accelerator pump cam is set in the 2 position. In many instances, many of the cams will not work properly in the 1 position because the initial delivery action is quite slow (causing engine hesitation), whereas those same cams will work very well once positioned in the 2 position. Always think in terms of having the least amount of fuel being injected into the engine as a starting point.

Warning! - ALL cams can be used with 50cc accelerator pumps, but only some can be used with 30cc accelerator pumps. The brown and yellow cams are 50cc pump specific. **The accelerator will jam open if they are used with 30cc accelerator pumps if the adjustment factor is correct.**

ACCELERATOR PUMP DISCHARGE NOZZLES

There are two main types of discharge nozzle for 4150, 4160 and 2300 Holley carburetors: the 'straight type' and the 'tube type'. Both work well. Discharge nozzles frequently get drilled out after original manufacture so, unless the discharge nozzles you have are all brand new, check each hole for size so that you know for sure that they are not oversized. This does mean you'll need access to a set of number drills, but if you're serious about tuning your carburetor yourself this sort of equipment is required and you will use the drills a lot. Insert the unfluted shanks of appropriate sized drills into the holes in the nozzle. Use the drills as 'go' and 'no go' gauges to ascertain the actual hole sizes.

There is a range of standard sizes indicated by the numbers stamped on the sides of discharge nozzles. The number 28, for example, indicates the diameter of the hole size in the discharge nozzle in thousands of an inch (i.e. - 0.028in). The range of actual fuel discharge hole sizes for the 'straight type' of discharge nozzle are as follows: 0.021in, 0.025in, 0.028in, 0.031in, 0.035in, 0.037in, 0.040in, 0.042in, 0.045in, 0.047in, 0.050in and 0.052in. The 'tube type' of discharge nozzle is available in 0.025in, 0.028in, 0.031in, 0.035in, 0.037in, 0.040in, 0.042in and 0.045in diameter hole sizes.

Another type of four way discharge nozzle that was available on 4160 type (660 CFM) carburetors under List Number 4224 and 4150 type (855 CFM) carburetors under List Number 3418-1, but they're not

BASIC CARBURETOR PARTS

'Straight type' left and right and a 'tube type' in the centre. A 'tube type' is shown in the centre which has 0.035in diameter discharge holes (denoted by the numbers '35' stamped on the side of it. The 'straight type' nozzle on the left has '31' stamped on it which demotes 0.031in diameter discharge nozzle holes in it and the 'straight type' nozzle on the right has 28 stamped on it which denotes 0.028in diameter discharge holes in it.

up through the circuit and in the nozzle from draining back on 'hanging ball' check valve fuel bowls. With the system working correctly, the discharge nozzle is loaded with fuel ready for instant action.

Note. Holley advises that when larger than 0.037inch discharge nozzles are used (that's larger than part number 121-237, 121-37 or 121-137 discharge nozzles), the alternative discharge nozzle screw (Holley part number 26-12) should always be used as it allows more fuel to pass around it than the standard screw. The screw can become the limiting factor to fuel flow (as opposed to the discharge

common these days as these were racing carburetors. They were called 'centre-shooter' carburetors and they were mechanical in action: the four holes discharging fuel simultaneously.

Yet another type of discharge nozzle was used on 4165 and 4175 model 'spread bore' Holley carburetors known as the 'anti pull over' type. These carburetors operate without a discharge needle valve, hence the discharge nozzle design difference to prevent siphoning. These 4165 and 4175 discharge nozzles can be fitted to 4150, 4160 and 2300 model carburetors, where applicable, and work very well. They are a drop in fit and are available in the following diameter hole sizes: 0.025in, 0.028in, 0.031in, 0.035in, 0.037in, 0.040in, 0.045in, 0.047in, 0.050in and 0.052in. A 0.025in diameter hole discharge nozzle has 25 stamped on it (number stamped either on the face of the diaphragm cover or the edge of the die casting), and the others are coded similarly in relation to the diameter of the holes drilled in them.

Keep spare new washers (gaskets) on hand for each discharge nozzle as, once the washers start to leak, all of the fuel will not come out of the discharge

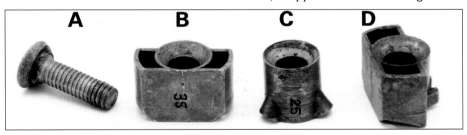

Two views of an 'anti pull over' type of discharge nozzle (B,D). The size comparison between a straight type (C) and the 'anti pull over' types. The securing screw (A) with the sealing washer stuck to it. Anti pull over-type of discharge nozzles from 4165 or 4175 models of Holley carburetor can also be used on 2300, 4150 and 4160 carburetors.

nozzle holes but rather will be sprayed in other directions from the damaged washer. Replace any washer that looks damaged to avoid failure in service.

Always check that a needle type check valve is fitted under each discharge nozzle. Failure to fit a needle type check valve will result in engine hesitation during acceleration. The reason the needle type check valve is there is to maintain fuel above the check valve, which in turn is above the level of the fuel bowl. If the needle valve is not fitted, the accelerator pump has to fill the accelerator pump circuit above the fuel level of the carburetor fuel bowl before fuel can be discharged from the nozzle. The check valve prevents the fuel that is pumped

nozzle holes) if you do not use this screw.

ACCELERATOR PUMP MECHANISM FUEL BOWL NON-RETURN VALVES

There are two main types of non-return valve in the accelerator pump mechanisms of these carburetors. They are the 'umbrella type' (two versions) and the 'hanging ball type,' the 'hanging ball' type being the most common. Both non-return valve types work well and, although there are slight differences in their individual operation, they don't really make any significant difference in the overall scheme of things.

The 'hanging ball type' valve

SPEEDPRO SERIES

Discharge nozzle for 4150, 4160 and 2300 carburetors comprises a needle type check valve (extreme left), a washer that goes between the carburetor body and the base of the discharge nozzle, the discharge nozzle (centre), a washer that fits between the top of the discharge nozzle and the underside of the retaining screw (retaining screw is on the extreme right). The parts are fitted to the carburetor in order of left to right.

works on the principle that with the steel ball resting on the small bar that limits its travel, fuel can drain into the accelerator pump and also any air that finds its way into the accelerator pump diaphragm can vent back into the fuel bowl. The ball must be free to move but not have too much up and down movement. The instant the accelerator pump is activated the movement of fuel lifts the steel ball and forces it against the seat so that the fuel in the diaphragm cannot go back into the fuel bowl where it came from. With the check valve shut off the fuel is displaced into the accelerator pump circuit and ultimately into the carburetor through the discharge nozzle. When the accelerator pump action stops and the diaphragm is returning to its normal position, the steel check ball drops back onto the bar and fuel can then flow into the diaphragm from the fuel bowl to replace the fuel that has been discharged into the engine. The valve ball is in the open position until the accelerator pump is activated. The ball must be able to move up or down 0.010-0.013in.

The 'umbrella type' valve performs the same function although it is closed except when taking fuel into the diaphragm. Once the accelerator pump is primed and full, the 'umbrella

Very common 'rubber umbrella'-type of check valve fitted into this 'side hung'-type fuel bowl.

type' check valve remains shut and only opens when the diaphragm has discharged fuel into the engine and is on its return stroke. The suction effect easily opens the 'umbrella type' valve and lets fuel into the diaphragm ready for the next accelerator pump discharge action. The discharge valve is shut except when taking in fuel on the return stroke of the diaphragm which discharges fuel to the discharge nozzle immediately the diaphragm moves (there is no steel check ball to move a against a seat and prevent fuel going back into the fuel bowl).

The advantage of the 'umbrella type' check valve is that it is

'Hanging ball'-type of check valve fitted to this 'side hung'-type fuel bowl.

permanently shut once fuel has passed into the diaphragm and is ready for instant action. The disadvantage is that air, if it gets into the accelerator pump diaphragm, cannot be vented back into the fuel bowl as it can with the 'hanging ball type' check valve, but this rarely a problem.

MAIN JETS

The main jet sizes start at 40 and go through to 100. The numbering system is not entirely related to the diameter of the holes in thousandths of an inch, although there is a fairly close relationship in the group of main jets from 40 to 66 which are, as it happens, approximately 0.040in to 0.066in in diameter. After number 66 main jet, however, the diameter of the holes starts to move away from the numbering system (they get larger than the number size stamped on the main jet). Main jet number 100 has a hole diameter of approximately 0.128in.

Anyone who has only ever worked with jet sizes ranging from 40 to 66 could be forgiven for thinking that the numerical listing and the hole diameter size in the main jet are always directly related. However, if you consult a Holley main jet chart you

BASIC CARBURETOR PARTS

Selection of main jets.

The Holley size number of a jet is stamped on the side of the main jet. This one is stamped '74' and has an approximately 0.081in diameter hole drilled through the middle of it.

quickly see that they are not: this is another aspect of carburetors which is frequently misunderstood.

Holley main jets are also numerically rated in relation to flow. There are counter drillings each side of the hole in the middle of the main jet and the depth of these holes, as opposed to the diameter, plays a significant part in determining the flow rating of the main jet. This means that drilling out the centre hole of the main jet to equal, say, the next Holley numerical sizing does not work. This is because the counter drillings will not necessarily be the same as the next sized main jet and, therefore, the flow

Here is a full list of Holley main jets giving the hole diameters of each in thousandths of an inch.

40 - 0.040in	71 - 0.076in
41 - 0.041in	72 - 0.079in
42 - 0.042in	73 - 0.079in
43 - 0.043in	74 - 0.081in
44 - 0.044in	75 - 0.082in
45 - 0.045in	76 - 0.084in
46 - 0.046in	77 - 0.086in
47 - 0.047in	78 - 0.089in
48 - 0.048in	79 - 0.091in
49 - 0.048in	80 - 0.093in
50 - 0.049in	81 - 0.093in
51 - 0.050in	82 - 0.093in
52 - 0.052in	83 - 0.094in
53 - 0.052in	84 - 0.099in
54 - 0.053in	85 - 0.100in
55 - 0.054in	86 - 0.101in
56 - 0.055in	87 - 0.103in
57 - 0.056in	88 - 0.104in
58 - 0.057in	89 - 0.104in
59 - 0.058in	90 - 0.104in
60 - 0.060in	91 - 0.105in
61 - 0.060in	92 - 0.105in
62 - 0.061in	93 - 0.105in
63 - 0.062in	94 - 0.108in
64 - 0.064in	95 - 0.118in
65 - 0.065in	96 - 0.118in
66 - 0.066in	97 - 0.125in
67 - 0.068in	98 - 0.125in
68 - 0.069in	99 - 0.125in
69 - 0.070in	100 - 0.128in
70 - 0.073in	

rate might be different from Holley specification.

Holley does not consider that main jets should be drilled out and recommends that only new main jets be used in the carburetors.

Frequently people drill out main jets just to find out what basic size of main jet is right for their engine. Then, once they have the size more or less finalised by testing, new main jets are substituted. If these new jets are not quite right, the next size (up or down) is bought new and tested until the right size is found. Clearly, when a new engine has been built, the jetting for it might have to be altered quite substantially, and to have to buy the complete range of jets to do this will become quite expensive. In these circumstances, drilling out main jets is a sensible option for testing purposes to find the best jet size. However, fitting correctly sized new main jets at the point of final testing is strongly recommended.

Whenever jets are drilled out, the two primary jets and the two secondary jets must be of the same part number. To drill a 54 main jet and a 72 main jet out to some new larger size, such 75, for example, is not recommended, even for testing purposes, as the two counter drilling holes of the jets will be quite different in drilled depth and, therefore, the fuel flows will definitely be different.

Caution! - Always use a screwdriver with a blade that fits the full width of the slot in the main jet as this will avoid 'burring'.

Note that when main jets are ordered the prefix '122-' must be placed in front of the number of the actual jet. So when a 75 jet is ordered, for example, you ask for part number 122-75 main jet.

Note that Holley has also made a range of what it calls 'close limit jets' for some carburetors which are interchangeable (the listing is in the *Holley Performance Parts Catalog*). These main jets have three numbers on them as opposed to two, and are classed as high accuracy main jets by Holley. These main jets are perfectly acceptable for all applications (the first two digits determine the main jet sizing).

SPEEDPRO SERIES

NEEDLES AND SEATS

There is a range of needles and seats to suit the various types of float bowl. 2300, 4150 and 4160 'side hung' non externally adjustable fuel level float bowls use part numbers 6-511, 6-516 and 6-510 needles and seats. Externally adjustable 'side hung' and 'centre pivot' fuel bowls for these carburetors, on the other hand, use part numbers 6-517, 6-508, 6-504, 6-501, 6-500, 6-502 and 6-515 needles and seats. In the past there have been other part numbers for needles and seats but this is the current listing at the time of writing.

The majority of Holley carburetors are equipped with 'Viton' tipped needles (rubber tipped). Steel needles are available for use with alternative fuels (fuels other than gasoline/petrol). Various sizes of needles and seats are available. If the seat size (in fact, the diameter of the hole in the seat) is too small the engine will run out of fuel at wide open throttle. Conversely, if the seat size is too large it can cause fluctuating fuel levels in the fuel bowls which will cause the engine fuelling to fluctuate between a correct mixture and a slightly rich one. The needle and its seat need to be the right size. The Holley recommended needle and seat size for the particular List Number carburetor is the starting point and, 95% of the time, will prove to be absolutely right for any application.

In the past Holley made the needles and seats so that the fuel flowed out of the needle and seat and into the fuel bowl through round holes. All current Holley made needles and seats have rectangular holes (which they call 'windows') for the fuel to flow out of the actual needle and seat assembly and into the fuel bowl.

Some non-genuine Holley replacement part kits contain needles and seats which have holes drilled in them (like the early Holley needles and seats, as opposed to the now usual Holley 'windows') and have a screwdriver slot in the top for float adjustment. The securing nut also has an O-ring on its underside to prevent fuel leakage when under pressure. This is quite a user-friendly adjustment system and is generally well thought out. However, because of the method of construction of these aftermarket seat assemblies they cannot be checked for seat hole size by inserting drill shanks. Refer back to the particular kit manufacturer for size details, or check the suitability of the needle and seat by road testing. Most are equivalent or nearly equivalent to a genuine Holley D needle and seat.

These aftermarket needles and seats are very acceptable for all general road going applications and many high-performance applications, but they can become a limitation in that they simply cannot pass enough fuel into the fuel bowl. They are also less prone to leaking fuel than the standard Holley ones when the washers are not in perfect condition. The genuine Holley adjustment system of a base washer, adjustment nut, washer and securing screw works very well provided the washers are in perfect condition, but it is without a doubt more 'finicky' than the aftermarket adjustment arrangement.

Viton-tipped steel needle and seat assembly.

If the carburetor runs out of fuel during wide open throttle work, change the needle and seat to a 'window' type. For competition work, always use genuine Holley parts (or aftermarket ones which have 'windows' in them)

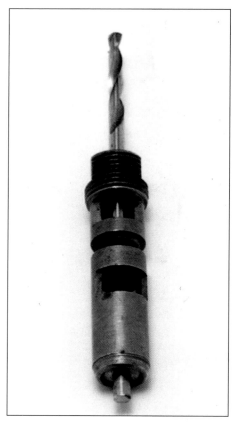

Drill shank fits neatly into the drilled hole of the seat and right down to the tip of the needle.

BASIC CARBURETOR PARTS

This needle and seat assembly has the capital letter 'D' stamped on the flat and has a black Viton-tipped needle and a 0.098in diameter holed seat in it.

Here is a list of currently available Holley needles and seats for the carburetors covered by this book.

D: 0.097-0.098in diameter seat & Viton tip needle

J: 0.097-0.101in diameter seat & Viton tip needle

H: 0.110in diameter seat & Viton tip needle

A: 0.097in diameter seat & steel needle

C: 0.110in diameter hole & steel needle

F: 0.120in diameter hole & steel needle

M: 0.130in diameter hole & steel needle

for maximum possible ease of fuel flow into the carburetor fuel bowl.

Always check to make sure that the O-ring fitted to the needle and seat is in good condition, and not perished. Keep on hand spare washers for the adjustment nut and the securing screw which holds the needle and seat in place in the fuel bowl.

The size of the needle and seat of a genuine Holley component is indicated by a letter stamped on one of the flats machined on the threaded section. The factory use a capital letter stamping to identify each needle and seat assembly. The D and J needles and seats are of very similar size and the main difference between them is that the D is a two piece assembly while the J is a swaged one piece assembly. Either way, the size of the needle and seat is suitably identified.

Sometimes the letter stampings are not all that clear. In these cases to find out what needle and seat sizes actually are, firstly look to see what type of needle is in the assembly: is it a Viton tipped needle (black or red conical rubber tip), or a needle made entirely out of steel? Secondly, the hole size in the body of a genuine Holley needle and seat assembly can always be measured by carefully inserting the shanks of various Number sized drill bits into it and determining the diameter on a 'go'/'no go' basis. Number drill bits can usually be bought from engineering supply companies.

If the needle and seat you have is an aftermarket copy of a genuine Holley late model 'window' type assembly, but it is not letter stamp marked, the checking procedure already described will allow you to find out what the seat size is. Unfortunately, aftermarket closed top, screwdriver slot type needle and seat assemblies cannot be measured by the method described so you'll have to refer to the relevant manufacturer for equivalent size data to genuine Holley parts.

PRIMARY METERING BLOCKS

All primary metering blocks have idle mixture adjustment screws, accelerator pump circuitry and power valves. Each primary metering block model is unique by way of various drillings for idle circuits and the fact that the power valve restriction channels are sized to suit the CFM ratings of the matching carburetors. The bigger the carburetor the larger the PVCR (power valve channel restriction) hole size. This last factor is one very good reason for making sure that the metering blocks you fit to the carburetor body you have are the right ones for that carburetor, or, at the very least, the CFM rating of the carburetor. Holley goes to a lot of trouble to match the various drillings in the metering blocks to the CFM rating of the carburetor and, if the carburetor is sized correctly there will be few problems. The range of PVCR hole diameters on the Universal Performance Holley carburetors starts at 0.038in for the smaller CFM four barrel carburetors and goes up to a maximum size of 0.072in on large CFM four barrel carburetors.

Some primary metering blocks for smaller CFM four barrel carburetors have an O-ring tube connection for the accelerator pump discharge from the metering block into the body of the carburetor (390-CFM four barrel). Only the specific design of metering block can be used on this type of carburetor body. Both components are identified as being correct for each

SPEEDPRO SERIES

Letter and numbers are stamped on the top edge of metering blocks. The significant characters on this example, 'L18502', are the List Numbers of the carburetor, making it easy to identify. The fact that the metering block has idle adjustment screws, the hole drilled and threaded to take a power valve and holes drilled in it for the accelerator pump circuit means that this is a primary metering block.

other by the List Numbers stamped on them.

There are some quite subtle differences between the various models of primary metering block besides the PVCR hole sizes. There are various other drillings in the metering blocks for the idling circuitry as well as air bleed holes for the main jet well. A particular set of these variable factors constitute the correct calibration of a particular metering block for a particular CFM carburetor. This is why, if the power valve channel restriction hole size at some stage proves to be too large for an engine application, fitting an aftermarket kit with changeable jets

In the three photos shown, the positions and sizes of the holes and the number of holes varies metering block to metering block. The holes at A are for the idle circuitry. Those at B are the main jet discharge passageways in the metering block which take the air/fuel mixture to the main discharge nozzles in the venturis. The small holes indicated at C are air bleed holes into the main well.

BASIC CARBURETOR PARTS

'O' ring and tube connection for accelerator pump circuit on the primary side of a 390CFM four barrel carburetor body (List Number 8007).

and has no idle adjustment screws, it is definitely designed to fit a mechanical secondary 'double pumper' carburetor. All secondary metering blocks have idle circuitry because all four barrel carburetors have secondary idling systems which work all of the time.

All 4150 series vacuum secondary four barrel carburetors have secondary metering blocks. A secondary metering block designed to be used on this type of carburetor does not have the accelerator pump passageway holes drilled in it, which is a clear giveaway as to what the intended use of the particular metering block is. Vacuum secondary carburetors do not have secondary accelerator pumps, only primary ones. So, if you have a metering block which has not been machined to take a power valve, or been drilled for accelerator pump operation and does not have idle adjustment screws it is definitely designed to fit a vacuum secondary 4150 carburetor.

All 4150 and 4160 carburetors which have secondary metering blocks with power valves fitted to them have the same PVCR holes size regime as primary metering blocks, i.e. the hole sizes range from 0.038 to 0.072in (from smaller four barrel carburetors to larger four barrel carburetors). Most four barrel carburetors do not have secondary power valves fitted or adjustable secondary idle circuitry.

All secondary metering blocks have progression from idle circuitry in them. **All** secondary metering blocks have cast in provision to take a power valve (but the holes are mostly not drilled and tapped to take a power valve, they are left blank). **All** secondary metering blocks have cast in provision for idle adjustment screws (but most are undrilled and left blank). **All** secondary metering blocks have idle circuitry in them as all four barrel carburetors have secondary idle circuits which are either adjustable or non adjustable: most have non-adjustable secondary idle circuitry.

Alternative 4150 mechanical secondary metering blocks
Some secondary metering blocks have curb idle adjustment screws as well as operative power valves and accelerator pump discharge circuitry, making them

is often the optimum way to reduce the hole size (as opposed to getting another metering block with smaller factory drilled holes in it). What can happen here is that while the power valve channel restriction aspect of the metering block will be corrected, the idle circuitry and the progression phase calibration will not be correct.

SECONDARY METERING BLOCKS

All 4150 series 'double pumper' or mechanical secondary four barrel carburetors have metering blocks and all of these metering blocks have accelerator pump circuitry drillings in them. So, if you have a metering block which has been drilled for accelerator pump operation but not been machined to take a power valve

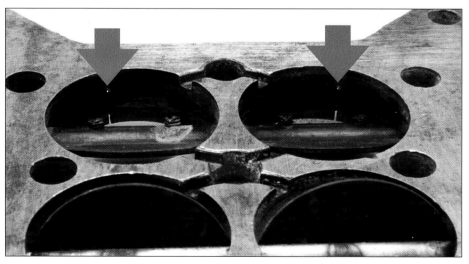

ALL four barrel 4150 and 4160 Holley four barrel carburetors have secondary idling circuitry and small holes in the two back barrels (shown by the arrows).

SPEEDPRO SERIES

the same as primary metering blocks. Secondary metering blocks like this are not common and are intended for large CFM mechanical secondary four barrel carburetors (such as 750-CFM List Number 9379, 830-CFM List Number 9381, 850-CFM List Number 9380 carburetors as well as 750-CFM List Number 9645 and 850-CFM List Number 9646 methanol calibrated carburetors, for example). These metering blocks will fit on the primary or secondary side of a four barrel carburetor.

Four barrel, 'four corner idle' carburetors could have identical or different metering blocks front and rear. The reason for this is to do with the PVCR hole sizes of the metering blocks which can give different fuel curves for different applications.

There are a considerable number of possibilities when it comes to identifying secondary metering blocks, but all can be categorised with a bit of thought. Essentially which metering block can be fitted to one of these carburetors depends on the body of the carburetor. There are two general shapes to the secondary side of a four barrel carburetor body.

The first is the flat back, as found on vacuum secondary carburetors which have a metering plate fitted to them (and no possibility of having a power valve equipped metering block fitted as there's no cast in recess for power valve clearance or drillings). A carburetor body like this could have a metering block fitted to it for easy main jet changes, but the metering block will have to be one that did not have the power valve hole drilled and tapped when it was made by Holley.

The second type of carburetor body has the cast in recess for power valve head clearance. However, even if the back of the carburetor body has the recess cast into it, it does not mean that the body has been drilled by Holley so that the power valve will actually work. If the actuating holes have been drilled in the cast in recess of the body (clearly visible), then this carburetor will take a power valve and it will be operative. If you have a secondary metering block which is drilled and tapped to take a power valve but don't want the power valve to operate, a plug (Holley part number 26-36) will have to be fitted.

METERING BLOCK VENT BAFFLES

Holley has made four different fuel bowl vents over the years which fit directly to all metering blocks. One is a 'metering body vent baffle' which is a thin brass pressing sandwiched between the fuel bowl gasket and the metering block. This was fitted as standard equipment. The second is a perforated baffle (Holley part number 26-39) which is made of thin brass or aluminum sheet and is also sandwiched between the fuel bowl gasket and the metering block. The third is the 'whistle' vent baffle (Holley part number 26-40), an injection moulded plastic item which fits into the metering block and which is held in place by a pin.

This third vent baffle type is of more interest than the other two for competition purposes. This baffle is very good at preventing fuel sloshing back into the primary metering block's vent aperture on acceleration. This is because the 'whistle' vent baffle opening is at the front of the fuel bowl and, while fuel can be sloshing around the metering block end of the 'whistle', it will not affect the running of the engine. If the vent hole in the metering block gets filled with fuel and effectively blocked off, the engine will not run correctly as there is no atmospheric pressure acting on the fuel in the fuel bowl. Fitting a 'whistle' vent baffle to the primary metering block is ideal for drag cars and can be beneficial for circuit racing cars too. In recent years, a fourth type of baffle has become available which is suitable for front or rear metering blocks. This one (black plastic) is a 'press in fit'. It's shorter than the 'whistle' type vent, but long enough to prevent the fuel that gets 'sloshed' around in the fuel bowl due to either acceleration or braking action, from covering the end/ends and preventing the engine from running correctly. These vent baffles are

Thin brass shim 'vent baffle' sandwiched between the metering block and the gasket.

BASIC CARBURETOR PARTS

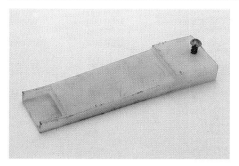

'Whistle' type vent baffle.

'Whistle' type vent fitted into a metering block.

Later short type plastic vent baffle.

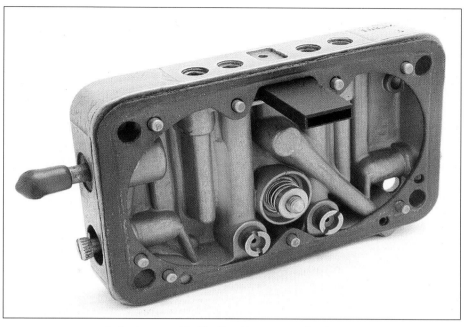

Later type vent baffle fitted into a metering block.

suitable for all kinds of competition and road use.

SECONDARY METERING PLATES

Secondary metering plates all have one or two numbers stamped in the middle of the flat face, and they are quite easy to identify from these numbers. Furthermore, all secondary metering plates have idle circuitry in them.

The *Holley Performance Parts Catalog* has a metering plate listing under the heading 'Holley secondary metering plates'. The list covers all 52 secondary metering plates by number and by full part number which means that prefix '134-' is placed in front of the actual number/numbers stamped on the metering plate. The sizes of the various holes as drilled in each metering plate are recorded in the Holley listing.

When dealing with metering plates, you need to be aware of the fact that there are six metering plates that share three single identification numbers. This is not a problem when buying new metering plates from Holley because they are sold packaged up by part number so the fact that two totally different metering plates can have the same identification number on them just doesn't matter. The problem occurs when used metering plates are being identified. If you look at the Holley chart there are two different metering plates sharing the numbers 3, 4 and 5 (that's component part numbers 134-3, 134-4, 134-5, 134-53, 134-55 and 134-54). As can be seen from the Holley metering plate chart 134-3, and 134-53 are number stamped 3, 134-4 and 134-54 are number stamped 4, 134-5 and 134-55 are number stamped 5. However, these six metering plates can be identified further by casting numbers. 134-3, 134-4 and 134-5 carry casting numbers 34R-5113B. 134-53, 134-54 and 134-55 carry casting numbers 34R-9716B. All of the other

SPEEDPRO SERIES

This metering plate has the number '34' stamped on it. The numbers are always upside down when the metering plate is screwed onto the carburetor body.

secondary metering plates can be ordered from the chart once you know exactly what metering plate you already have.

The diagram accompanying the 'Holley secondary metering plates' (in the *Holley Performance Parts Catalog*) shows the two holes identified as 'A' and 'B': 'A' being the main jet holes or 'secondary main metering restrictions', and 'B' being the 'secondary idle feed restrictions'. The sizes of the holes at 'A' and 'B' are what makes the difference between the various metering plates which are otherwise identical.

If the secondary metering plate you have is secondhand, it's a good idea to check the sizes of the holes at 'A' and 'B' just to make sure that they have not been drilled out. Check the 'Holley secondary metering plates' listing to find out what the hole sizes should be in the metering plate that you have. Then check with the shanks of Number sized drill bits what the hole

45 metering plates are individually numbered and are completely identifiable by these numbers.

The *Holley Performance Parts Catalog* lists the part numbers of the metering plates fitted to all of the 4160 series carburetors listed in the Numerical Listing, with all numbers prefixed '134-' followed by one or two digits. The *Holley Performance Parts Catalog* has another listing in it under 'performance carburetor fuel bowls, metering blocks, throttle body assemblies'. What is not so straightforward in the 'secondary metering block' listing is that the 134- one and two digit part numbers are, in fact, metering plates and **not** metering blocks as implied by the list name. All 134- numbers followed by three digits are metering blocks.

The 'Holley Secondary Metering Plates' listing (in the *Holley Performance Parts Catalog*) gives the size of the secondary idle feed holes, the secondary main metering hole sizes, the part numbers for ordering purposes and the numbers stamped on the metering plate for identification purposes. The secondary metering plates are listed in lean to rich order in the chart (listed hole sizes getting progressively larger). Richer or leaner

Holes at 'A' and 'A' are main jet holes while the holes at 'B' and 'B' are secondary idle feed holes.

BASIC CARBURETOR PARTS

This panel shows a list of Holley secondary metering plate hole sizes and relevant identification numbers.

'A'	'B'	ID	'A'	'B'	ID
0.052	0.026	7	0.076	0.029	43
0.052	0.029	34	0.076	0.031	12
0.055	0.026	3	0.076	0.035	3
0.059	0.026	4	0.076	0.040	28
0.059	0.029	32	0.078	0.029	38
0.059	0.035	40	0.078	0.040	52
0.063	0.026	5	0.079	0.031	11
0.064	0.028	18	0.079	0.035	24
0.064	0.029	30	0.081	0.029	44
0.064	0.031	13	0.081	0.033	49
0.064	0.043	33	0.081	0.040	21
0.067	0.026	8	0.081	0.052	31
0.067	0.028	23	0.081	0.063	29
0.067	0.029	16	0.082	0.031	46
0.067	0.031	9	0.086	0.043	25
0.067	0.035	36	0.089	0.031	47
0.070	0.026	6	0.089	0.037	5
0.070	0.028	19	0.089	0.040	27
0.070	0.031	20	0.089	0.043	26
0.070	0.033	41	0.093	0.040	4
0.071	0.029	35	0.094	0.070	15
0.073	0.029	39	0.096	0.031	50
0.073	0.031	37	0.096	0.040	45
0.073	0.040	17	0.098	0.070	14
0.076	0.026	10	0.113	0.026	42
0.076	0.028	22			

Primary barrel progression (from idle) slots are shown arrowed 'A'. All barrels have these slots. Idle holes shown arrowed 'B' are for adjustable curb idle. The holes at 'B' are under the edge of the butterflies when in the idle position.

sizes actually are. Unknowingly using a drilled out metering plate can be the cause of major tuning complications.

Caution! Do not fit the fluted part of the drill bit into the holes in the metering plate as they can mark the surfaces of the hole.

Holley doesn't recommend drilling out any of these metering holes, but it **is** a useful practise when secondary jetting is being sorted out during testing. A number drill set and a pin chuck is needed for this.

It's not possible to sensibly tune and alter the jetting of a Holley carburetor unless you truly know what components are in it. The absolute identification of parts like metering plates, and then checking the integrity of those parts, is vital to getting the engine tuned correctly in a reasonable amount of time. If a metering plate has been drilled out and you don't know it, a huge amount of time can be wasted wondering what is going wrong with the secondary mixture.

If there is no listing of your carburetor in the Numerical Listing, e-mail Holley to establish what metering plate should be on the particular List Number carburetor you have.

In the *Holley Performance Parts Catalog* there is a cross-reference chart for the conversion of main jet sizes (as fitted in metering blocks) to metering plates, or when metering plates are changed for metering blocks. Few people will be changing from a secondary metering block to a metering plate system, but the conversion can be used in either circumstance.

The heading in the *Holley Performance Parts Catalog* under 'Holley Secondary Metering Plates' mentions secondary idle feed restrictions. This is really a reference to the progression circuit (which feeds air/fuel mixture to the small slots that

SPEEDPRO SERIES

Arrows at A are primary barrel curb idle holes, arrows at B are primary barrel progression holes and slots, and the holes/slots at C are for secondary idle and progression.

This carburetor body is 'flat backed' and only needs a single gasket.

Metering plate has been fitted to this carburetor body using the one main gasket.

This carburetor body has a cast-in recess to take a power valve.

are machined in the primary and secondary barrels of all of these Holley carburetors), which is in operation from the near fully closed position through the first 10-15 degrees of butterfly opening. This is nothing to do with curb idle or engine idle circuitry which is adjustable via the idle mixture adjustment screws. In fact, it is to do with the first phase of acceleration.

All four barrel carburetor throttle body and shaft assemblies have secondary idle progression slots and secondary curb idle circuitry. Secondary curb idle circuitry can be adjustable or non-adjustable: most being non-adjustable. On non-adjustable carburetors, idle holes are directly under the progression slots, which means that one air/fuel mixture feed hole supplies the secondary idle and the progression feed slot.

Both secondary metering plates and metering blocks have circuitry that feeds air/fuel mixture to these slots and (very small) curb idle holes. The slots only come into full operation when the secondary butterflies move from the closed position and sweep past the slot, thereby subjecting it to engine vacuum. The feed holes are already flowing fuel to the idle holes which are on the inlet manifold side of the butterflies (downstream of the butterflies). These progression slots are to assist with smooth engine operation when the butterflies are opened from the closed or near closed position. The air/fuel mixture drawn from the slots as the butterflies open smoothes out the throttle response just after the butterfly's initial opening and before the secondary barrels start using the main jet air/fuel discharge circuit.

FITTING METERING PLATES TO CARBURETOR BODIES

Secondary metering plates are fitted in one of two ways: One way sees the metering plate screwed directly to the

BASIC CARBURETOR PARTS

a carburetor body that has cast in provision for a power valve. In this situation there will be the usual large gasket that fits between the metering plate and the back face of the carburetor body, then a thin metal support plate (in the shape of the metering plate), then a gasket (also in the shape of the metering plate), and then, finally, the metering plate itself. This arrangement ensures the metering plate is pulled up against a solid surface. Without the support plate being fitted to a carburetor body which has the cast in provision to take a power valve, the metering plate would have only the gasket as its backing face in its middle section. Clearly this is not correct at all and, while it might work for a time, the gasket will eventually fail through distortion. **Caution!** - The support plate and gasket **must** be used on a carburetor body which has the cast in provision for a power valve. Many Holley carburetor bodies are like this.

FUEL BOWL, METERING BLOCK, METERING PLATE GASKETS

There is quite an array of different gaskets for these Holley carburetors. To find out exactly which gaskets you'll require for fuel bowl, metering block and metering plate, you'll need the List Number of your particular carburetor model. If your carburetor is not listed, take the whole carburetor to a Holley parts supplier and get the gaskets which match the individual components and make a note of the part numbers of each gasket for future reference.

Gasket requirements (using Holley part numbers for each gasket), can be summarised as follows, but there are some exceptions:

All 2300 carburetors use 108-40 body

The metering plate is fitted onto the carburetor body with the components in the following order (closest to the back of the carburetor body goes on first)

carburetor body, with a single gasket interposed. This method applies to completely 'flat backed' carburetor bodies. It does not need the steel support plate or the similar shaped gasket, although if you do fit these two items as extras it won't detrimentally affect performance.

The second fitting method is when a metering plate is used on

SPEEDPRO SERIES

Part of the range of vacuum secondary springs. Springs are of different lengths, tensions and colours.

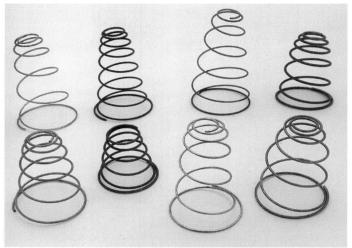

Assortment of vacuum secondary diaphragm springs.

Diaphragm spring is fitted into the housing as shown in the photo.

to throttle body and shaft assembly gaskets.
All 2300, 4150 and 4160 carburetors use 108-33 fuel bowl to metering block gaskets.
All 2300 and 4150 carburetors use 108-29 primary metering block to body gaskets.
450-780-CFM 4160 carburetors use 108-29 primary metering block to body gaskets.
390-CFM 4160 barrel carburetors use 108-31 primary metering block to body gaskets.
All 4160 four barrel carburetors use 108-30 secondary body to fuel bowl gaskets.
Some vacuum secondary 4160 carburetors use 108-27 metering plate gaskets.
All 4150 carburetors use 108-33 fuel bowl to metering block gaskets.
All 4150 carburetors use 108-29 secondary metering block to body gaskets.
4150 and 4160 carburetors use a range of carburetor body to throttle body and shaft assembly gaskets, go by the diameter of the throttle bores of the body.

BASIC CARBURETOR PARTS

Check ball 'A' goes in the hole as shown by the arrow.

VACUUM SECONDARY DIAPHRAGM SPRINGS

There is a range of 10 color coded diaphragm springs for 4150/4160 vacuum secondary carburetors, and they are available singularly or in a kit.

There are two aspects to these springs. The first is that the 'softer' the spring is at the installed height within the diaphragm housing, the quicker the diaphragm will open, although no spring can even start to open until there is enough vacuum being generated to lift the check ball in the mechanism. Always check to make sure that the steel check ball is fitted in the diaphragm housing.

The other feature of vacuum springs is the rate at which they allow the secondary throttle shaft to open once the vacuum threshold has been reached. The lower the spring rate, the quicker the barrels will open. The correct opening of the secondaries is governed by spring heights and spring rates.

The ideal situation is to have the spring fitted that allows the engine to accelerate the car as fast as possible. This is quite easy to achieve by timing how long it takes to cover a measured distance while accelerating, or timing how long it takes for the car to be accelerated to a certain speed as indicated by the speedometer.

The colors of the springs are: white, yellow, red, purple, green, pink, orange, plain unpainted, brown and black.

The ideal spring for your application is best found by trial and error. Settle for the firmest spring that gives the quickest acceleration. Correct spring choice will make sure that the primary barrels are near to reaching their maximum airflow capability before the secondaries start to open.

If the carburetor is sized correctly for the engine, the 'softer' springs (white, yellow, red, pink or green), will usually work better than the 'harder' springs (brown and black). Consider the black and brown springs to be generally more suitable for use with large capacity engines (7000cc and above).

The vacuum secondary diaphragm spring fits on a raised section on the cover housing. The vacuum secondary operation check ball 'A' goes in the hole indicated by the arrow in the accompanying picture. The check ball should be fitted for almost every application.

VACUUM SECONDARY DIAPHRAGMS

There are three sizes of diaphragm listed under part numbers 135-2, 135-3 and 135-4 for the vacuum secondary four barrel Holley carburetors. Each model of four barrel vacuum secondary carburetor has to have the right diaphragm of the three fitted.

The differences between the secondary diaphragms is to do with the length of their shafts and is nothing to do with the neoprene diaphragm itself (they're all identical). The reason for the different shaft lengths is to do with the linkage plate on the end of the secondary spindle and the position on this linkage plate of the pivot. The three different linkage plates available give three arcs of travel and, consequently, three slightly different opening rates for the secondary barrels.

Details of the vacuum secondary diaphragms for popular four barrel carburetors is as follows:

Holley part number 135-2: shaft approximately 1.865in long
Holley part number 135-3: shaft approximately 1.965in long
Holley part number 135-4: shaft approximately 2.045in long

Quoting the List Number of your carburetor to a Holley parts agent will identify exactly which part number diaphragm a particular carburetor

Linkage plates vary in location pin position.

57

should have. If the carburetor you have is not listed in the *Holley Performance Parts Catalog* under the 'Secondary Diaphragm' listing, take the carburetor to a Holley parts supplier and find out by 'trial and error fitting' which of the three diaphragms fits (it only takes a minute).

Caution! Diaphragms are very easily damaged when the cover screws are being removed or fitted. The problem is the threads of the four small screws can catch the diaphragm and start to turn it which, if it's not noticed, will frequently ruin the diaphragm. The solution to this problem is to lubricate the threads of the screws and the edges of the diaphragm where the screws pass with petroleum jelly. To reduce the prospect of ruining a diaphragm when stripping a carburetor, undo two diagonally opposite screws fully without touching the other two screws.

Now turn each of the two remaining cover securing screws out a half turn and then spray a penetrant (CRC or WD40) on to each screw head. In a few seconds the penetrant will have gone down the thread of each screw, and the screws can be removed without them catching the diaphragm.

To check the seal of the diaphragm and its condition, the housing has to be dummy assembled with the check ball removed. With the unit assembled in this fashion, the shaft of the diaphragm is pushed in until it bottoms out. Now place a finger firmly over the air inlet hole in the housing and release the diaphragm shaft. The shaft should not move. If it moves the seal is not effective and the diaphragm is damaged, not fitted correctly (and now likely irreparably damaged for that reason) or there is something wrong with the two metal parts of the housing. If the diaphragm proves to be properly sealed, carefully dismantle the assembly, re-fit the check ball and carefully reassemble it all over again.

Damaging the vacuum secondary diaphragm when the housing cover is removed to change the diaphragm spring is something that can be completely avoided. Plastic housings are available which have a removable top centre section which allows you to get at and change the spring without having to remove the top housing. These housings are very worthwhile if you intend to experiment with vacuum secondary diaphragm springs because the risk of damaging the diaphragm is removed, because you don't remove the top housing cover, and the time it takes to change the spring is significantly reduced (1 minute).

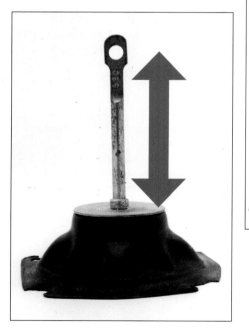

The measurement is taken from the base of the shaft to the centre of the hole.

The three vacuum secondary diaphragm and shaft assemblies suitable for four barrel Holley carburetors.

Chapter 5

Tuning 2300 two barrel carburetors & the primary barrels of 4150 & 4160 four barrel carburetors

THE RELATIONSHIP BETWEEN TWO AND FOUR BARREL CARBURETORS

There's a relationship between 2300 two barrel and 4150 and 4160 four barrel Holley carburetors. The primary barrels of the 4150 and 4160 are the same as the two barrels of the 350 and 500CFM 2300. Looked at another way, because the four barrel carburetors came first, the 2300 series 350 and 500CFM two barrel carburetors are the front half, or primary section, of either of these four barrel carburetors.

Four barrel carburetors are by far the most popular and widely used performance carburetors for V8 engines and have been for many years. In spite of this factor, many people still do not seem to be able to get their carburetors tuned correctly. Compared to two barrel Holley carburetors, the four barrel units are more complicated to tune, of course, which can lead to frustration and an engine that does not perform as it should and which consumes too much fuel.

Unfortunately, it's not uncommon to find 4150 and 4160 four barrel Holley equipped engines misfiring, 'bogging down' after curves and hesitating from a standing start at all manner of competition events. This level of tuning is almost unbelievable because, if such problems were directly attributable to the carburetor, all could be easily eliminated if a reasonably straightforward tuning procedure was followed. If the sizing (CFM) choice is correct, it's not the carburetor causing the problems, it's the way it has been set-up ...

The easiest way to tune four barrel Holley carburetors is to disable/disconnect the secondary barrels and tune the primary barrels first as a separate entity. Then reconnect the secondary barrels and tune them as separate entities. The primary barrels are set correctly in the first instance and not touched again. Following this procedure will prevent 95% of the tuning problems that people experience with the Holley four barrel carburetors.

FUEL SUPPLY

It's important to address the issue of fuel supply before you start to tune an engine. Otherwise, you could end up trying to deal with problems that you perceive to be tuning-related, but which are, in fact, down to fuel supply.

These carburetors need fuel pressure in the vicinity of 3.5 to 6.5psi/2.4-4.5kPa, and enough volume to ensure that the fuel pressure is maintained under all circumstances. To this end a few things have to be attended to. As a general rule, the outlet of any fuel tank used needs to have a minimum 5/16in outside diameter steel pipe coming off it, with a inside diameter of 1/4in (some tanks have larger piping, like 3/8in, for example). This will match the usual unions that most fuel pumps and

SPEEDPRO SERIES

Holley carburetors come with, as well as most stock equipment mechanical fuel pumps.

Note that the stock fitting fuel unions for most Holley fuel bowls have a 1/4in inside diameter. The tube that takes fuel to the rear fuel bowl on a four barrel carburetor fitted with 'side hung' fuel bowls has a 3/16in inside diameter.

The petrol/gasoline resistant hose used to fit onto the tank outlet will be 5/16in inside diameter. That hosing then goes to the primary side of the mechanical/electric fuel pump, and must have as few bends in it as possible. If there are any bends, they must have as large a radius as can be accommodated in the installation. The hosing must not be clamped to the body/chassis in any way that reduces its effective diameter. It has to be clamped firmly in position to the underside of the car, or inside the car, of course, but there are ways of doing this without 'crimping' the hosing excessively. Hosing must also be kept well away from any heat source, like the exhaust pipes, for example. If hosing passes through a bulkhead, the hole needs to have a grommet fitted to prevent the steel or aluminium cutting into the hosing over time. Two small, neat fitting, stainless steel 'Jubilee' type clamps should always be used on the fuel tank connection, if possible (it usually is).

Fitting a fuel filter at or near the fuel tank will prevent dirt clogging the fuel pump. Such filters need to be well protected against any possible damage, and be replaced at regular intervals (annually, for example). Use small, neat fitting, stainless steel 'Jubilee' type clamps at each connection (as opposed to the 1/2in wide heater hose type clamp which is perhaps more commonly available). On balance, it's often better to fit the fuel filter away from the engine (at the rear, for example, where the fuel tank would be on a front-engined car). The engine installation dictates to a degree what you can and can't do.

With the fuel going through the mechanical fuel pump at the front of the engine, it comes out of the pump under pressure. If you haven't fitted a fuel filter before the pump, you need to fit one now. It's usual to continue with 5/16in inside diameter fuel hosing up from the pump, and to some convenient point where the fuel filter can placed in-line where it's least likely to get damaged, be away from a heat source, yet be convenient to get at. The filter and the hosing need to be clamped to keep them in position.

Whether the fuel filter is fitted before or after the mechanical fuel pump, the fuel line now does one of two things. It either goes to the fuel inlet of a two barrel carburetor or to the front 'side hung' fuel bowl of a four barrel carburetor equipped with this type of fuel bowl; or it divides by one of two means to feed fuel to each 'centre pivot' fuel bowl of a four barrel carburetor so equipped. It is at this point (before the fuel goes into the carburetor bowl/s), that a 'take off' for a fuel pressure gauge needs to be placed. The hose that comes off this can either be permanently plumbed into the instrument panel, or fitted temporarily. On a road car, for instance, once you've established that the system works as required, there's no need to monitor the fuel pressure unless you think there may be a problem at some later stage. Racing cars, on the other hand, almost always have a fuel pressure gauge fitted as a permanent part of the instrumentation.

Because the stock Holley fuel filters get blocked over time, and are quite expensive to replace, it's often best to remove them when their service life is over, and rely on an external Holley or aftermarket fuel filter instead. The stock Holley fuel filters either fit into the carburetor fuel union, or behind the fuel inlet, or they form part of the fuel inlet. Like all filters, they get blocked, and because of their positions it's easy to forget that they are there. Blockages can reduce the amount of fuel going into the fuel bowls and, since the filters are placed in the carburetor bowl/s after the fuel pressure gauge, your fuel pressure gauge could be showing normal fuel pressure when it's not actually the case. It's better to remove these fuel filters and prevent the possibility of this happening.

Most stock original equipment mechanical fuel pumps supply between 3.5-4.5psi/2.4-3.1kPa, which, although less than the maximum permissible, is enough for most applications, including racing. Many aftermarket mechanical fuel pumps will pump between 4.5-5.5psi/3.1-3.8kPa and sometimes more, but, provided it doesn't exceed 6.5psi/4.5kPa, there is no need to use a regulator. Most mechanical pumps can't pump more than 5.5psi/3.8kPa. It's most unlikely that a large capacity mechanical fuel pump will not be able to supply enough fuel to a naturally aspirated engine.

When the fuel level of the fuel bowl/s drops too much the engine will run lean and 'backfire' through the carburetor (the first warning sign of what's going on). This commonly happens when an engine is at maximum sustained rpm, such as on a long straight. **Note.** Don't persist in running the engine at high rpm, find out what's causing the problem!

Holley makes a range electric fuel pumps, some of which require a regulator to reduce the fuel pressure to your requirements. In the vast majority of instances, 4.5-6.5psi/3.1-4.5kPa will

TUNING 2300, 4160 & 4150 CARBS

500CFM two barrel 2300 carburetor.

be sufficient for all naturally-aspirated engines. These pumps can keep up with the demand of virtually any engine. They are usually fitted near to the fuel tank, and are often positioned for gravity feed to their primary side.

Other electric fuel pumps are available (from Facet, for example). Provided the pump is sized correctly for the capacity of the engine it is to feed fuel to, these pumps will supply an engine at about 4-4.5psi/2.7-3.1kPa and they don't need a regulator. If there is a problem with volume and pressure you can always fit two pumps side by side. That's one pipe coming out of the fuel tank, dividing neatly into two pipes, going through the two pumps, and then going back into one line via neat bifucated jointing.

INTRODUCTION TO TUNING & ADJUSTMENT PROCEDURES

This chapter covers the general tuning procedure for both the 2300 two barrel carburetors and the primary side of all 4150 and 4160 Holley four barrel carburetors (the latter are identical to the 2300 two barrel carburetor).

Please note that the photos used in this chapter are a mix of two barrel and four barrel carburetors. The differences between the carburetor models should be ignored unless the relevant text or caption says otherwise.

Tuning the secondary side of 4150 and 4160 carburetors is covered in chapter 6.

Important! The tuning method described in this chapter is presented in a progressive sequence which must be adhered to if you want to get the very best from your Holley carburetor or carburetors.

Warning! Every time a carburetor is refitted, check that **all** fuel line connections are secure and that the fuel lines are well away from exhaust pipes.

Caution! When tuning engines equipped with four barrel carburetors, but with the secondary barrels' throttle mechanism disabled, testing the engine using sustained high rpm is **not** recommended. What is being advocated here is to check the acceleration of the engine to the point of, but not beyond, the maximum air flow of the two barrels. If an engine (any engine and any capacity) accelerates well with the primary barrels of a four barrel carburetor up to 5500rpm, for example, and then 'flattens' (reaches the peak CFM flow of the two barrels), do not persist in trying to take the engine's rpm higher. If the engine clearly reaches a peak at 5500rpm use 5300rpm after the first time test.

The danger is that with the secondary barrels closed off, a high inlet manifold vacuum could be generated at high rpm. This could close the power valve, weakening the mixture, with possibly disastrous results if the engine speed is sustained for more than a very brief period.

If, during testing, a vacuum gauge is temporarily fitted in full view of the driver or rolling road operator the amount of manifold vacuum can be constantly monitored. If the amount of vacuum being generated at high rpm is exceeding the rating of the power valve, the engine will have to be shut down and a higher numbered power valve fitted, or only be tested in this manner up to an rpm point just before the power valve is going to close. The general and sensible procedure is to

SPEEDPRO SERIES

fit a power valve of a higher value (a 3.5Hg one over a 2.5Hg one, for example), than the maximum vacuum seen on the gauge. Note that this might not prove to be the best power valve to have fitted to the carburetor for the application but at this point you just can't have the engine going into vacuum. The requirement is to initially fit a power valve .5Hg to 1.0Hg above the maximum vacuum gauge reading seen.

Don't ignore the possibility of high inlet manifold vacuum problems at high rpm if you intend to take the rpm to dizzy heights on the primary two barrels only of a four barrel carburetor when a low rated power valve (2.5Hg, for example), is fitted into the carburetor. The same goes for a two barrel equipped engine turning high rpm when the CFM rating of the carburetor is being exceeded. The power valve just has to be rated higher than the maximum vacuum possible. If this factor is not compatible with the engine's idle requirements (a long duration camshaft, 295-300 plus degrees, for example), the engine's rpm will have to be limited so that the inlet manifold never goes into a state of vacuum higher than the power valve rating.

FUEL LEVEL ADJUSTMENT ('CENTRE PIVOT' & 'SIDE HUNG' FUEL BOWLS)

To set the float level of a 'centre pivot' fuel bowl well enough to start the engine without risk of severe flooding proceed as follows. With the fuel bowl off the carburetor, place it upside down on a bench so that the needle and seat is in the shut off position. Now measure the distance between the top of the float and the top of the inside of the fuel bowl (in the middle of the fuel bowl). The measurement can be taken with a Vernier calliper or with the shank of a $3/8$in diameter drill bit: the size of the gap required is $3/8$in.

To set the float level of a 'side hung' fuel bowl well enough to allow starting the engine without the risk of flooding proceed as follows. With the fuel bowl off the carburetor, place it upside down on a bench so that the needle and seat is in shut off position. The brass floats and Duracon floats need to be positioned so that they sit symmetrically in-line with the underside surface of the fuel bowl. A Nitrophyl float needs to be positioned so that the top surface of the float is in-line with the underside surface of the fuel bowl.

On the top of both types of fuel bowl there is a large flat-headed screw. This screw holds everything in place and clamps the needle and seat in position. The nut underneath this screw is the float level adjustment nut (it serves no other purpose). Between the underside of the nut and the fuel bowl there is a fibre washer and, between the top of the nut and the underside of the large flat-headed securing screw, there is another fibre washer. **Warning!** It pays to have spare washers on hand because if one or both of these washers are damaged while the float level is being adjusted, fuel leakage is very likely. These small fibre washers (available from Holley parts suppliers) get damaged when the nut and the securing screw are turned.

To adjust the float level accurately proceed as follows: Undo the large set screw with a suitable screwdriver a $1/3$ of a turn out from the fully tightened position. With the screwdriver held in this position, use an open ended wrench (spanner) to turn the hexagon nut underneath the large flat headed screw. Turning the adjustment nut clockwise (that's when looking down on it when you are facing the front of the carburetor) lowers the float level and anti-clockwise raises the float

Holley fuel pressure gauge (left), electric pump (centre), and regulator (right).

TUNING 2300, 4160 & 4150 CARBS

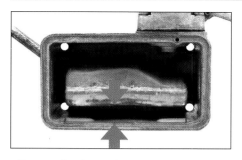

'Centre pivot'-type fuel bowl showing the approximate position of the float when the fuel bowl is upside down with the needle and seat in the shut off position. The distance between the float and the fuel bowl needs to be approximately 3/8in.

'Side hung'-type fuel bowl, which has a small brass float, showing the approximate position of the float when the fuel bowl is upside down and the needle and seat in the shut off position. Similar shaped Nitrophyl and Duracon floats are positioned similarly.

'Side hung'-type fuel bowl, with a long brass float, showing the approximate position of the float when the fuel bowl is upside down and the needle and seat in the shut off position.

level. Keeping the retaining screw in close proximity to the adjusting nut as it is turned ensures that the nut is always in good mesh with the top of the needle and seat. If the needle and seat is to be raised (adjusting nut turned anti-clockwise), the retaining screw will need to be wound clockwise an equal amount to maintain the slight clearance between the underside of the retaining screw and the adjustment nut. Conversely, if the float level is being lowered the needle and seat is going to be wound clockwise, this means that the retaining screw will need to be turned anticlockwise an equal amount to maintain the slight clearance between the underside of the retaining screw and the adjusting nut. It takes a bit of practice to do this, but it does ensure that the adjustment nut always has the maximum contact with the top of the needle and seat.

When adjusting the float level the needle and seat should turn **freely** in the float bowl thread for the full range of potential adjustment. If the brass needle and seat threaded body cannot be turned easily in the float bowl the float bowl's internal thread could be damaged, or the needle and seat thread could be damaged, or both. Either way this problem needs to be addressed and put right before adjustment is attempted. It is quite common for the thread in the float bowl to need to have a 'cleaning' tap run through it. This procedure can be done by a garage or an engineering workshop and will take a mechanic or an engineer about a minute, provided they have the right tap (3/8in - 32 threads per inch plug tap). If the needle and seat thread is badly damaged, replacement is required, though a suitable die nut could be used to repair the thread.

The reason that this thread problem is important is that if the needle and seat become tight the high torque applied to the nut can seriously damage the top of the needle and seat and also the nut itself can become damaged. Not only can the needle and seat become jammed in the float bowl (meaning the float level cannot be adjusted to the right height), but the needle and seat can be extremely difficult to remove too. **Caution!** If the adjustment nut becomes difficult to turn, stop immediately and remove the float bowl to find out exactly what the problem is. The needle and seat **must** freely wind down into the float bowl until the top of the needle and seat is level with the top of the float bowl. This is well past the point of maximum adjustment, so, if the needle and seat become unthreaded a small pair of long nosed pliers can be used to turn the needle and seat out sufficiently so that the nut can be fitted back on to the top of the needle and seat so that they can easily be wound out of the fuel bowl. The top of the needle and seat has two machined flats on it which the adjustment nut straddles and this is how the needle and seat is turned in or out. It is, of course, only on well used float bowls that this problem is likely to be encountered.

These externally adjustable floats are raised or lowered to set the float 'shut off' height (the point at which the fuel supply is switched off) by turning the adjustment nut clockwise to lower the float level or anti-clockwise to raise it. Fine adjustment of the fuel level is only made once there is fuel in the bowls, of course. Both types of fuel bowl have a sight plug on one side and it is via the sight plug hole that accurate float levels are set. The sight plug is removed for float level setting purposes. In the majority of instances the sight plugs of two and four barrel carburetors are on the same side of the carburetor, irrespective of what

SPEEDPRO SERIES

'Centre pivot'-type fuel bowl float adjustment componentry (right). The needle and seat (arrowed) positioned in the fuel bowl. Large diameter fibre washer on the right is first on over the brass threaded portion of the needle and seat, followed by the hexagon nut, followed by the small diameter fibre washer: all is held in place by the securing screw on the left.

type of fuel bowl is fitted. Note that the latest type of Holley carburetors have a non-removable clear plastic sight plug instead of the earlier removable brass screw-in plug, and on these carburetors the fuel level is viewed only through the sight plug. Adjustments of the float are made up or down with the fuel being used by the engine if the float level is too high. If the floats have been adjusted as suggested before being fitted to the carburettor, only minor float adjustment will be required.

If the car has an electric fuel pump, the fuel level can be adjusted without even starting the engine. With the car on **level** ground the fuel pump is switched on and the level of the fuel checked by seeing if fuel comes out of the sight plug hole once the fuel bowl is full. If no fuel comes out of the sight plug hole the fuel level is too low. If fuel pours out of the sight plug hole the fuel level is too high. If fuel just dribbles out of the sight plug hole, the fuel level (float setting) is correct. Replace the sight plug after setting the float level.

If the car has a mechanical fuel pump the engine will have to be turned over on the starter to fill the bowl up

Fuel bowl securing screw and washers.

Float level adjusting mechanism assembled (arrowed).

64

TUNING 2300, 4160 & 4150 CARBS

but not actually started. Disconnect the ignition. Then, with the car on **level** ground crank the engine over with the sight plug removed and fill the fuel bowl up. If the fuel starts to flow out of the sight plug hole, stop cranking the engine and lower the fuel level via the float adjustment. If the fuel does not flow out of the sight plug hole adjust the float until it does. The requirement is for fuel to dribble out of the sight plug hole which means that it's just level with the bottom of the hole. Replace the sight plug after setting the float level.

PRIMARY BARREL IDLE SPEED AND IDLE MIXTURE

On the throttle arm side of the carburetor there is a throttle adjustment screw which, if turned clockwise, for example, increases the amount of butterfly opening. Turning the screw anti-clockwise reduces the amount of butterfly opening.

To adjust this screw so you can start the engine, turn it anti-clockwise until the end of the screw is no longer in contact with the throttle actuating lever. Then, with some hand pressure on the actuating arm to shut the throttle, turn the screw clockwise and continue to do so until you feel the throttle arm just start to move. Stop turning the idle speed adjustment screw and note the position of the slot in the screw's head. Turn the screw $^3/_4$ of a turn in the same direction from this position. This will be a sufficiently accurate setting to start the engine and have it idle at a reasonable speed in order to make fine adjustments.

On each side of the metering block there is an adjustment screw which adjusts the air fuel mixture ratio of the barrel on that side of the carburetor. Prior to starting the engine turn the two adjustment screws clockwise so that they are

Sight plug and its position (arrowed) on a primary 'centre pivot'-type fuel bowl.

fully seated (lightly seated) and then turn them back $1^1/_2$ turns. **Caution!** Excessive screwdriver pressure to seat the idle mixture adjustment screws is neither required nor desirable as the carburetor body can be damaged by this means, making accurate idle mixture adjustments difficult.

SECONDARY BUTTERFLY IDLE ADJUSTMENT SCREWS (4150 & 4160)

Before fitting any four barrel carburetor to an engine, check what position the secondary butterflies are in. There is an adjustment screw on the underside of the carburetors that needs to be adjusted correctly. It's quite possible to have the rear butterflies open too much, letting too much air into the engine. In this case you won't be able to get the engine's idle speed down to a sensible level, even if the primary butterfly adjusting screw is completely wound off. It's also quite possible to have the secondary butterflies shut so firmly against the throttle bores of the baseplate that they can 'stick' shut (vacuum secondary carburetor only). This only happens if the adjustment screw has been wound off completely and is not touching the secondary throttle spindle link arm at all.

In most cases these very small adjustment screws will never have been adjusted since the carburetor left the factory and they will often be firmly jammed. The solution to this problem is to apply WD40 or CRC and then limited heat to the throttle body and shaft assembly which will expand the aluminum and loosen the fit between the adjustment screw and the aluminum. **Caution!** The screwdriver used to turn the adjustment screw must be a **perfect** fit in the slot: if it isn't, and the screw is jammed in position, the slot could end up being so badly damaged that it will have to be drilled out. Avoid this situation at all costs ...

What needs to be taken into consideration with regard to the secondary throttle spindle adjustment is the fact that if the primary butterflies have to be open too much to effect a decent idle, there can be progression problems from idle. What happens is that with the primary throttle butterflies open more than they should to allow a good idle, the edges of the butterflies are not in close enough proximity to the idle transfer slots of the throttle bores. This lack of proximity means that insufficient air/fuel mixture is drawn out of the idle transfer slots as the edges of the butterflies sweep past them as the throttle is opened. Hesitation is quite possible in these circumstances and the fault can be difficult to track down and fix unless you know about this potential problem. No normal amount of accelerator pump action will cover this problem.

This has led to holes being drilled in the butterflies of primary barrels so that they can remain more or less

SPEEDPRO SERIES

Engine idle speed adjustment screw (arrowed at 'A') on a two barrel carburetor.

Left-hand venturi idle mixture adjustment screw (arrowed) for 2300 carburetors and the primary venturis of 4150 and 4160 carburetors.

Right-hand venturi idle mixture adjustment screw (arrowed) for 2300 carburetors and the primary venturis of 4150 and 4160 carburetors.

shut, the air for idling purposes going through a small factory-drilled hole in each primary butterfly. It's a good idea, and it works.

The requirement for **all** four barrel carburetors is to have the primary butterflies being near to fully shut off and flowing about 50% of the air required for idling purposes, and the secondary barrels also flowing about 50% of the air required for idling purposes. This means having near to equal amounts of throttle or butterfly opening of the primary and secondary butterflies. It doesn't have to be exact but it has to be near to this requirement.

To adjust the secondary butterflies correctly the primary and secondary adjustment screws are wound out until they are not touching the linkage contact points, which means that the butterflies are in firm contact with the throttle bores. The primary and the secondary butterfly adjustment screws are in turn wound in until they touch the linkage arm contact points and then wound in a further $1/2$ turn each. Next the carburetor is fitted to the

Engine idle speed adjusting screw (arrowed at 'A') on a four barrel carburetor.

TUNING 2300, 4160 & 4150 CARBS

Position of the secondary butterfly adjustment screw arrowed at 'A'.

remove the adjustment screw from the throttle body and shaft assembly, and screw it in from the top. You will then be able to adjust the secondary butterfly position without taking the carburetor off each time. While this sounds good in theory, it's not so good in practice because, if you look at the standard screw you will notice that the screwdriver slot has been splayed out. This has been done by Holley so that the screw is hard to turn in the threads and is effectively self locking. The solution to this small problem is to cut a new slot into the radiused end of engine and the engine started to see what the idle speed is. If the idle speed is what you want the adjustments are correct.

If the engine will not idle fast enough, increase the secondary barrel opening by 1/8 of a turn and see what happens then. If it is not enough, do it again, and then turn the primary barrel adjusting screw in a further 1/8 of a turn.

If the engine is idling too fast, reduce the primary throttle adjusting screw setting by 1/8 of a turn. If this isn't enough reduce the secondary adjustment screws setting. It's easier to make each adjustment with the carburetor off the engine.

The very acceptable range of primary idle screw inward adjustment is between 3/8 to 5/8 of a turn, with 1/4 to 3/8 of a turn being about the minimum ever possible/desirable. This whole adjustment process may take a bit of time because the carburetor will have to come off to get at the secondary barrels' adjustment screw for each incremental adjustment. In most instances, however, it only requires the carburetor to come off once or twice.

A possible alternative here is to

Holes in your butterflies?

One technical point that is of interest with some two and four barrel Holley carburetors is the fact that some have butterflies with a small hole drilled in them. The 500 CFM two barrel carburetor, for example, has an 0.085inch diameter hole in each butterfly. These holes have been drilled in the butterflies to effect a reasonable idle speed without having to have the actual butterflies cracked opened by more than the smallest amount. What this does is allow the edges of each butterfly adjacent to the throttle bores (on the same side as the holes are) to sweep past very close to the idle transfer slot machined in the side of the throttle bores. The reason for doing this has nothing really to do with the idle, but more to do with throttle (butterfly) opening progression. The closer the edges of the butterflies are to the idle transfer slots, the smoother the progression phase of the throttle opening is going to be (more mixture is drawn into the carburetor barrels). This progession phase occurs between the instant the butterfly moves off from the curb idle position and for the first 1/8inch of its edge travel. If there were no holes in the butterfly, the butterflies would have to be open more than they are with the holes drilled in them to allow sufficient air into the engine. The effect of sweeping by would be reduced because the edges of the butterflies are all ready past the start of the idle transfer slot and moving away from the actual throttle bore as acceleration begins: the possible result is a lean progression mixture.

67

SPEEDPRO SERIES

This adjustment screw has been turned around and is now on the top side of the throttle body and shaft assembly plate. Once adjusted correctly a locking agent can be used to permanently lock the threads.

the primary butterflies to 'sweep' past, as much as possible, the full length of the 'idle transfer slots'. It's a vital adjustment.

Disabling the secondary barrels

With the idle speed set as just described, the secondary barrels must be prevented from opening. This can be accomplished by using thin gauge steel wire (0.047inch/1.2mm/18 gauge, for example), bound very tightly around the bottom of the diaphragm shaft and the throttle spindle linkage arm, and the onto a fixed point so that the butterflies cannot move. It's not necessary to undo the diaphragm shaft from the secondary throttle spindle.

There is some variation in the linkage systems by which the secondary barrels of mechanical carburetors are actuated, but there are two very common ones which use link arms. These two linkage arms are quite easy to undo and remove. With

the adjustment screw so that the screw can be turned from that end. Note that these screws are hardened and so a slot cannot be cut with a hacksaw or file, however, this isn't a problem if you have a small high speed grinder and a small parting wheel.

Simply taking the screw out and turning it around is not really an option because the screw will not self lock and will move if you run the engine, losing the setting. An alternative to slotting the original is to fit the screw from the wrong side, so to speak, and, after the secondary butterflies have been adjusted to the required setting, put a few drops of thread locking agent onto the exposed threads of the adjustment screw. The thread lock will run down into the aluminum of the throttle body and shaft assembly and permanently lock the screw.

A dramatic gain in engine acceleration (the complete lack of hesitation) can be the result of making sure that the primary throttle is near shut at idle, thus allowing the edges of

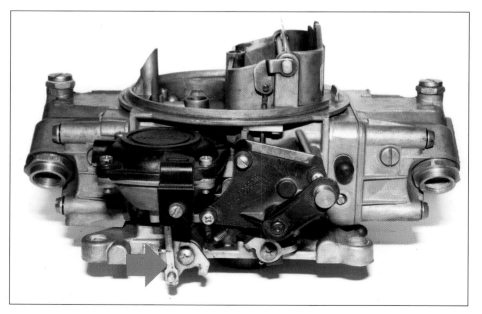

The vacuum secondary diaphragm shaft must not be allowed to move from this position. Wire can wrapped around the linkage arm (arrowed "A") and then around the choke mechanism or the diaphragm housing, for example, to stop movement.

TUNING 2300, 4160 & 4150 CARBS

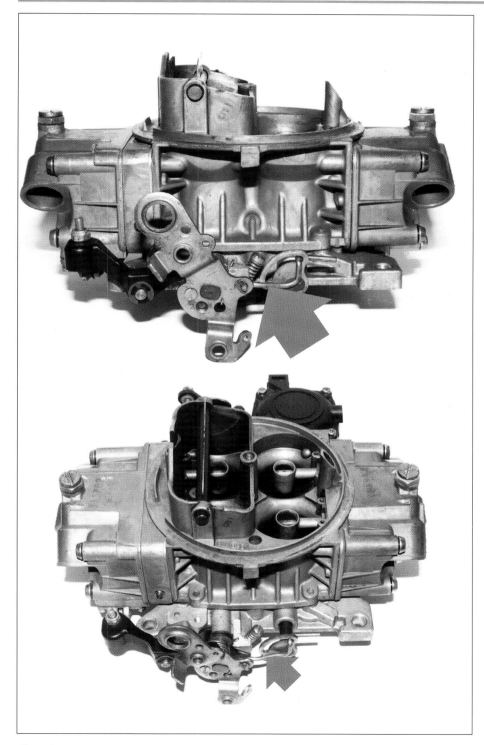

Two views of the throttle linkage limiting arm (arrowed) which can be removed by un-doing the split pin. The secondary linkage arm can then be wired closed.

the link arm removed the secondary spindle can be lock wired shut using the hole/slot normally occupied by the link arm.

The secondary butterfly spindle of all carburetors must not be allowed to move at all so the binding, using steel or copper wire, must be firm. The wire used to secure the linkage must be well away from the primary spindle actuating arm so that it can't interfere with the primary barrel's normal throttle operation.

In all cases where four barrel carburetors are being used the engine must be able to run correctly throughout the rpm range on the primary two barrels alone. Obviously this does not mean the engine will turn maximum possible rpm but, nevertheless, whatever rpm the engine is capable of turning to, the mixture strength has to be right. Don't be tempted to undo the wire holding the secondary barrels firmly shut to allow the two rear barrels to open to cure a problem somewhere in the rpm range. It's acceptable to expect an engine not to perform as well as it will with all four barrels working correctly but, realistically, engines go far better on the primary barrels alone than most people imagine. In fact, in many instances, engines go better using well tuned primary barrels alone than they will with an imperfectly tuned four barrel carburetor.

IDLE SPEED AND IDLE MIXTURE SCREW ADJUSTMENTS

From this point on the tuning procedure for 2300 carburetors and the primary barrels of 4150 and 4160 carburetors is identical (assuming the secondary butterflies of 4150 and 4160 carburetors are locked closed as previously described).

Initially, adjust the two idle

SPEEDPRO SERIES

The link arm (A) on this mechanical secondary/'double pumper' carburetor must be removed and the right hand link arm (B) bound securely shut using the slot in the link arm (C).

The link arm (A) is removed by removing the circlip (B) and slipping it out of the slot (C).

mixture adjustment screws to 1½ full turns out from the lightly but fully turned in position: this will ensure that there is enough idle mixture to start the engine. The number of turns and part turns of these two idle mixture adjusting screws must be accurate as later adjustments are made based on the initial settings.

The next initial setting adjustment is that of the accelerator pump mechanism (see 'Accelerator pump adjustment' section later in this chapter). The reason for adjusting the accelerator pump lever at this point is to ensure that the accelerator pump is working as near to full capacity as possible so that it can be used to inject fuel into the engine to facilitate starting and to keep the engine running until it has warmed up. Until the other settings have been narrowed down and running improved, using the accelerator pump in this manner can be very useful.

With the idle screw adjustment set, the accelerator pump lever can be adjusted. Adjustment of the accelerator pump must be done in this order because every time the idle speed adjustment screw is altered (throttle spindle position altered) the accelerator pump lever arm position must be reset to compensate. Failure to do this will result in a less than ideal accelerator pump action. With fuel in the bowl, all lash (play) needs to be taken out of the mechanism between the lever and the underside of the adjustment screw: they must be in contact but not firmly so. This is quite a fine adjustment and it needs to be absolutely right for instant accelerator pump response.

With the adjustments detailed so far the engine can be started. **Warning!** Check that **all** fuel line connections are secure, that the fuel lines are well away from exhaust pipes, and that all vacuum hoses are either connected or blanked off. With the carburetor full of fuel, the accelerator pump primed (work the throttle a few times) most engines will start more or less immediately.

If you are having trouble starting

TUNING 2300, 4160 & 4150 CARBS

the engine with these carburetor settings try an 'engine start' spray. These proprietary products contain ether/gasoline distillate mixture or pure petrol mixture in them and will start any engine after spraying the contents into the carburetor venturis. A 3-5 second burst into each venturi with the throttle open a bit is all it takes. **Warning!** Replace the air cleaner before starting the engine in case of a backfire.

With the engine started, allow it to warm up for a couple of minutes. Using the choke should not really be necessary to start a Holley carburetor equipped engine, unless it is a very cold day, but use it if you prefer. A $1/2$ to $3/4$ pump of the throttle pedal is frequently enough to get an engine to start and a couple of extra 'dabs' of the throttle pedal to inject a bit more fuel into the engine via the accelerator pump is usually enough to keep it running until the engine has warmed up enough to run on the idle circuit alone. Very cold climates might require the use of the choke on a regular basis.

Once a Holley equipped engine has been running for 10-15 seconds it will usually continue to run and, after about three minutes, be idling evenly enough to be able to start making adjustments to the carburetor.

The idle speed can usually be adjusted once the engine has been running for about two minutes, and certainly after four minutes when it will have attained the normal operating temperature. The idle speed for your individual application will depend on the type of camshaft fitted to the engine. The range of acceptable idle speeds for all engines is usually between 700rpm and 1500rpm. It is no use having an engine idling too slow for the type of camshaft fitted as this will lead to acceleration phase problems. You will have to be realistic here. The following is a rough guide:

Idling speeds
Standard camshaft: 700rpm
270 degree (smooth idle) camshaft: 900rpm
280 degree (lumpy idle) camshaft: 1000rpm
290 degree (rough idle) camshaft: 1100rpm
300 degree (very rough idle) racing camshaft: 1300rpm
300 degrees plus (very rough idle): up to 1500rpm

Essentially, idle speed should be set at a reasonably smooth pace. An engine set to run at too fast an idle speed often tends to 'run on' when switched off, especially when hot.

To finely adjust the two idle mixture adjustment screws (from the initial $1 1/2$ full turns out position) to obtain optimum idle smoothness, turn them both in $1/8$ of a turn and see how the engine responds. If the engine sounds 'lighter' the mixture was rich and is now leaner and further adjustments at an $1/8$ of a turn (in) at a time in are required. If the engine starts to misfire at odd intervals the mixture has become slightly weak and you know you have turned the idle screws in too far. In this case turn the idle mixture adjustment screws back out $1/8$ of turn to the previous good running position. At this point turn both idle adjustment screws back in to the lightly but fully seated position, accurately counting the number of turns and part turns required to do this. Now turn both screws back out to the best setting, on the basis of smoothness, found so far.

Note that it is usual for Holley carburetors to require an idle mixture screw adjustment of between $1/2$ a turn out from the lightly but fully seated position as a minimum (small capacity engine) to $1 3/4$ turns out as a maximum (large capacity engine) and seldom ever more than this. It is also quite usual for each idle mixture adjustment screw to be set with an identical 'turns out setting.'

To check each idle screw for best possible individual setting, adjust one screw in $1/8$ of a turn. Wait 10 seconds after adjusting the screw to see if there is any improvement in the engine's idle smoothness. If the engine idle gets rough turn the screw back out $1/8$ of a turn to the previous setting as the mixture has gone lean. Now turn that same screw a further $1/8$ of a turn out and see if that makes a difference to the idle speed smoothness. If it does make a difference turn the screw a further $1/8$ of a turn out and expect the idle to roughen at this point. Turn the screw back in to the middle position between when the idle mixture went lean from this present position (which is where the idle mixture went rich). This is the final individualised best setting for this side of the carburetor. Now go through the same process with the other idle mixture adjustment screw.

In most instances the idle mixture adjusting screws will not respond to this individual adjustment and will need to remain set equally. In other instances, however, one idle mixture adjusting screw will require a slightly different setting from the other to achieve optimum idle smoothness. This can take some experimenting to achieve. Once obtained, the settings will be set for the engine combination and it is advisable to make a note of them for future reference.

It is usual, and almost always quite possible, to get a Holley-equipped V8 engine to idle really smoothly with an extremely regular beat to it and with a 'lightness' of sound which you just know is right. Over-rich idle mixtures on the other hand cause an engine to have a 'heavy' sound. Setting the two idle mixture adjustment screws in the central position between being too lean and too rich usually achieves an excellent idle. The exception to this rule are engines with extremely long duration camshafts where slight idle mixture over-richness might have to be used.

At this point the engine will be idling at a reasonable speed and the idle will be as smooth as you can get it. There are no other adjustments to make to the carburetor on the basis of idle speed smoothness except, perhaps, to check the fuel level by removing the sight plug and seeing if the fuel level is still stable after the period of engine running that has occurred. Reset the fuel level (as described earlier) if there is not enough or too much fuel in the fuel bowl.

With the engine switched off, check that, with the accelerator pedal pushed right to the floor, the butterflies are vertical when you look down the venturis. This is one factor that is frequently the cause of poor engine performance. Any racing car should have the wide open throttle setting checked before any race meeting and on race day. It doesn't take a minute.

Note that with the engine's idle speed set and the two idle mixture screws set at the point of optimum engine idle smoothness, the accelerator pump must be readjusted so that there is no lash (play) in the mechanism. Any amount of lash means a delay in accelerator pump action.

ACCELERATOR PUMP ADJUSTMENT

For immediate fuel delivery by any accelerator pump to the engine, the diaphragm must be activated in unison with any throttle movement from the idle position. This is achieved by setting the idle speed of the engine in

Accelerator pump lever arm is arrowed at 'A' and adjustment screw is arrowed at 'B.' There must be no appreciable gap at 'C.' The lever arm 'A' MUST have sideways movement, but no up and down lash (freeplay).

the first instance and then checking and adjusting the accelerator pump actuating lever system for zero lash (play). The ideal is an accelerator pump lever which is in contact with the accelerator pump cam and the actuating lever of the fuel pump diaphragm housing with no perceptible free play (up or down movement) but having perceptible sideways play of the diaphragm lever. This ideal is achieved by adjusting all of the up and down movement out of the lever mechanism, but not so much so that there is no sideways play of this one lever. This is quite a fine adjustment. However, when the mechanism is adjusted like this, the instant the throttle is moved so the accelerator pump diaphragm will be moving and pumping fuel.

Don't be tempted to set the mechanism up so tightly that the pump arm has moved from its at rest position: this will mean the arc of potential travel has been reduced and therefore the pump 'shot' will be smaller.

Important! Every time the idle speed is adjusted, the accelerator pump mechanism must be readjusted if the ideal accelerator pump action is to be maintained. The reason the engine's idle speed has to be set first is because the accelerator pump cam height varies in relation to throttle arm position. This means that whenever the idle speed is increased, or decreased, the accelerator pump mechanism has to be reset so that the zero clearance is maintained.

Warning! If the accelerator pump lever is badly out of adjustment and well into its travel, the accelerator pump lever can end up being jammed against the accelerator pump cam, jamming the throttle in the wide open position. This situation will arise the very first time the accelerator pedal is fully depressed. **Always check (with the engine switched off),**

TUNING 2300, 4160 & 4150 CARBS

that the accelerator pedal can be depressed to the floor safely after any adjustment has been made to the accelerator pump mechanism. If the throttle jams open when the car is underway the consequences could be terrible. This potential problem is the reason for Holley's recommended 0.015in clearance between the diaphragm lever and the pump operating lever at full throttle.

The accelerator pump mechanism should be set as described so far before the car is road tested. In most instances the engine will accelerate correctly in the unloaded state with the stock accelerator pump cam and discharge nozzle as recommended in the Numerical Listing. However, this may change when the engine is under load. You won't know if the accelerator pump 'shot' amount is correct until you try the car on the road.

In many instances the standard discharge nozzle will be the right one, as will the accelerator pump cam. However, in just as many instances, the standard parts will not be absolutely ideal. Holley can only set a carburetor specification to meet a limited number of situations. A slightly rich accelerator pump shot is not ideal for accelerating a car, nor is a slightly lean shot. The slightly rich shot can fool you into thinking that it is right because there is no hesitation, whereas, in fact, the car is accelerating slower than it might because the engine has to burn off the excessive mixture before it can make ideal power. Some testing is required to narrow down whether the accelerator pump shot really is ideal or not. If the standard set-up proves to be absolutely correct after testing then all well and good, but you are not going to know this unless you test thoroughly. If the standard set-up is to your liking, with no hesitation or excessive black smoke

The accelerator pump cam can be fitted in one of two positions on the throttle lever pressing.

from the exhaust pipe under maximum loading acceleration, then leave it alone as it is very probably correct. If the engine does not accelerate cleanly follow the steps in the 'Accelerator pump cams' subsection.

Accelerator pump cams

A further aspect of accelerator pump tuning and adjustment is the fact that the plastic accelerator pump cams have two positions built in. The throttle arm has two holes in it to accommodate alternative cam positioning. With the cam fitted in position 1 the accelerator pump 'shot' will be smaller than when that same cam is fitted in the 2 position. What is vital here, of course, is that when the cam position is changed the accelerator pump arm to lever relationship changes pretty dramatically.

The range of 9 cams coupled with the 1 and 2 hole positions amounts to 18 possible individual accelerator pump actions. Some cams give near identical total amounts of fuel delivery per 10 pump actions, but all cams have a different action/fuel delivery amount per degree of throttle spindle rotation, from the idle position through to about 40 degrees of throttle spindle/butterfly opening. Since this is the vital difference between all of the 9 cam actions for most tuning applications, listing cams on the basis of full shot delivery is not always that relevant or useful. Optimum tuning may involve trying a few cams richer or leaner than those recommended by Holley in both positions to find the very best combination.

An engine that has the right amount of fuel will accelerate cleanly without any hesitation, while an engine that has an excessive shot of fuel fed to it will not hesitate but will not accelerate quite as well. Too much fuel will be evidenced by a large puff of black smoke from the exhaust. When the accelerator pump 'shot' is correct there will usually be a slight puff of

SPEEDPRO SERIES

A Cams in order of maximum fuel shot delivery rated over 10 strokes of the accelerator pump	B Cams in order of fuel delivery for the first 40 degrees of throttle spindle movement from idle
1 - black	1 - pink
2 - pink	2 - black
3 - white	3 - red
4 - red	4 - white
5 - green	5 - green
6 - orange	6 - orange
7 - light blue	7 - brown
8 - brown	8 - light blue
9 - yellow	9 - yellow

The pin drill chuck on the left will take all of the small bits in this metric drill bit set. The drills start at 0.3mm/0.0118in and go to 1.6mm/0.063in in 0.05mm/0.0019in increments. This is small enough for most Holley carburetor drilling work, and these sets are not expensive. Metric drills are commonly available in 0.1mm/0.004in increments and (less common) in 0.05mm/0.002in increments. Imperial 'number drill' sets are available (though more expensive) and generally increase in 0.001in/0.025mm increments from 0.014in/0.35mm - 'number drill' 80 - up to 0.043in/1.075mm - number drill 57. From number drill 56 (0.045in/1.125mm) up to about number drill 35, which will cover most Holley carburetor componentry, the increments vary between 0.002-0.003in.

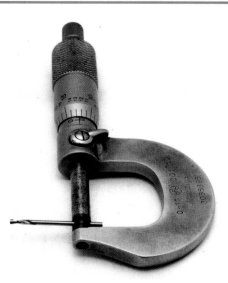

All small drill bits have to be measured using a micrometer on their shanks to ascertain their precise diameter. While drill bits might well be stored in a plastic case, they may not necessarily have been replaced in their correct positions in the case. Measuring drill bits before use is the best way to prevent errors.

black smoke or, in some cases, nothing at all.

Correct shot fuel volume will be clearly visible with a free revving engine (in neutral). In the free revving state the engine must be able to rev quickly without hesitation and with a minimum of black smoke being emitted from the exhaust. This is a good test and certainly the first to be used as a guide to getting the accelerator pump 'shot' setting right for your particular engine. If the engine cannot be accelerated quickly and cleanly in the free revving state, it will certainly not accelerate the vehicle under load. By the same token, just because an engine can be accelerated cleanly in a free revving state, does not mean that it will do the same when under load.

As an initial guide to finding the right accelerator pump cam and cam position (1 or 2), seek a hesitation free acceleration phase with the engine in the free revving state from idle speed through to 4000rpm. Most V8 engines will need cams set at the 2 position for the slight extra fuel delivery that this position will normally supply. Go through the whole cam range until the best combination, or combinations, are found. Even if the best combinations still have an amount of hesitation, find them and eliminate the ones that are definitely not suitable. Make a note of the cams that work and with what hole positions, then work the through the nozzle size options as these two factors are interrelated.

In order to do the testing necessary to find the best cam/hole combination, fit one of the following sized discharge nozzles appropriate for your engine:

3.5 to 4.2 litre V8: start with a 21 or 25
4.2 to 5.0 litre V8: start with a 25 or 28
5.0 litre plus V8s: start with a 28

The cams can be categorised to a certain extent, but the listing is quite loose because the cam actions are not *pro rata* from idle to the full shot position (some cams supply more fuel initially, while others pump progressively more fuel as the throttle is opened). The Holley listing is based on the cam's maximum fuel delivery capability, which is not in any way a representation of the cam's individual action. The list (A) is of the cams in order of 'full shot' fuel delivery as rated by Holley and does not take into account the rate of fuel delivery per degree of throttle spindle rotation from the idle position. Listing (A) is frequently more suitable for tuning racing engines where the full shot delivery size often really does matter. The second listing (B) is almost always more useful for obtaining optimum acceleration phase engine performance on road going and mildly tuned engines where the fuel shot delivery for the first 40 degrees of throttle spindle movement is often very critical. Both listings have their place in the overall scheme of things.

TUNING 2300, 4160 & 4150 CARBS

The rating of the nine cams is on the basis of the rate/amount of fuel delivery over the first 30 to 40 degrees of throttle spindle movement from the idle position. The list (B) is an approximate representation of each cams action in order of the least shot size to the maximum shot size. The changing from the number 1 hole position to the number 2 hole position will increase the shot size by a proportion of both charts.

Warning: Note that yellow and brown cams are for use with 50cc fuel pumps only. If these cams are used in conjunction with a 30cc fuel pump, the throttle will very likely jam in the fully open position the very first time the throttle is opened. The consequences could be disastrous.

There's always one cam/hole combination which, coupled with one of the range of accelerator pump discharge nozzle sizes, will provide the correct accelerator shot size. Your aim should be to use the smallest fuel shot possible to get the engine to accelerate cleanly without excessive black smoke from the exhaust.

Accelerator pump discharge nozzles

With the right amount of accelerator pump volume (30cc or 50cc), the right cam set in the right hole, and the accelerator pump set correctly, all that remains is to correctly size the discharge nozzle. Unfortunately, for testing purposes, having the full range of nozzles is more or less essential, as is being prepared on odd occasions to buy a smaller nozzle than you know is needed and drill the two holes out 0.001in at a time to end up with the perfect hole size for your particular engine.

Note that for drilling out the calibrated holes in Holley carburetor components you'll need accurately sized small drill bits held in a 'pin drill chuck'. These chucks are available from specialist hand tool supply companies and often come in sets of four sizes to cover the necessary range of drill bit sizes (usually up to $1/8$in or 0.125in). The drills are operated by hand.

Caution! - Unless you are completely confident of your engineering ability, avoid drilling components out and work within the confines of stock/standard components which will always take you very close to ideal.

The objective of this book is not to send readers down the carburetor engineering track, but rather to tune the particular carburetor you have by substituting standard readily available parts. There are, in fact, very few instances where the correct mix of standard parts will not work correctly and provide close to optimal carburation. However, the individual sizing (re-drilling) of discharge nozzles can mean that a more precise amount of fuel is discharged into a particular engine under hard acceleration.

Note that the smaller the discharge nozzle, the better the stream of fuel into the venturis. The obvious limitation here being that the discharge nozzle has to be large enough to flow the amount of fuel required by the engine. If the discharge nozzle is too small it could be the cause of hesitation, and simply changing to a larger size will usually cure the problem (if the discharge nozzle size was the problem). It is also feasible that, no matter what size the discharge nozzle is, if the accelerator pump is not working correctly, or is too small, the engine will hesitate under full load acceleration. The pump **must** work correctly, be of sufficient capacity and have the right rate of delivery.

It is very easy to 'drown' the engine with excess fuel from the accelerator pump without obvious hesitation, but no engine will accelerate well while it's burning off excess fuel. It may take time to achieve the optimum result, but it is always well worth the time and effort to get the accelerator pump shot size absolutely correct for a given application.

Of course, the jump between size 25 and 28 discharge nozzles, for example, can be more than is required. A 0.026in (number drill 71) might be perfect. In this instance, you could drill a size 25 discharge nozzle's holes out 0.001in once, try it, drill it out again 0.001in to 0.027in (0.7mm drill) and try it again, or you could get two 25 discharge nozzles and drill one 0.001in oversize and the other 0.002in oversize. It isn't usually this critical but many mechanics do go to this amount of trouble to get it exactly right.

At these small drill sizes, a combination of number drill and small metric drills can be required. The available metric drills that suit discharge nozzles, for example, are the 0.55mm/0.0215in, 0.60mm/0.0234in, 0.65mm/0.0254in, 0.70mm/0.0276in, 0.75mm/0.0297in, 0.80mm/0.0312in, 0.85mm/0.0337in, 0.90mm/0.0353in ones which, clearly, are not small enough to make minute adjustments. The number drills in the same diameter area are number 75 at 0.021in, number 74 at 0.0225in, number 73 at 0.024in, number 72 at 0.025in, number 71 at 0.026in, number 70 at 0.028in, number 69 at 0.0292in, number 68 at 0.031in, number 67 at 0.032in and number 65 at 0.035in. If you intend to go down the route of drilling Holley carburetor componentry, contact a specialist engineering concern about obtaining the exact drill sizes you are going to need for your application. There is no need to buy complete sets of expensive drills.

SPEEDPRO SERIES

The aim is to use the smallest discharge nozzle hole sizes, which still allow hesitation and black smoke-free acceleration under full load (the black smoke being caused by excessive fuel being injected into the engine). There are 9 individual accelerator pump cams, with two positions for each, and 12 different stock discharge nozzle sizes. This gives a very large number of possible combinations. To narrow the range, start with the recommended discharge nozzles and accelerator pump cams listed in the Numerical Listing for your carburetor, and work plus and minus around these recommendations. The majority of applications will require nozzle sizes between 21 and 35.

There are always going to be instances, however, where it proves difficult to get rid of hesitation. Likely causes are insufficient spark advance, a very long duration camshaft (over 295-300 degrees) and an idle speed that is too slow. It's also possible to go through the relevant range of stock discharge nozzle sizes and still end up with hesitation. If this happens, use the best nozzle size you've found so far and go back through the accelerator pump cam/hole combination testing procedure again. You're trying to ascertain whether one cam supplies a more appropriate amount of fuel during the acceleration phase than another. Make sure the accelerator pump mechanism is adjusted correctly after each change of cam/hole.

MAIN JETS

With the standard Holley recommended main jets and power valve for the carburetor List Number, the mains fuelling system needs to be checked to make sure that it is correct for the particular engine's state of tune. Selecting main jets in the first instance revolves around achieving optimum engine performance yet avoiding a lean or rich mixture anywhere in the fuel curve.

Setting up the main jet system is not always as straightforward as it might seem. There are two elements to take into consideration with a standard carburetor. One, the effect of the extra fuel supplied by the power valve when it is open, and two, the mixture supply when the power valve comes into operation. These are two very significant aspects of tuning these Holley carburetors because, if wrong, the engine will not perform at its best, yet it is easy to get them wrong and not know it ... What often happens is that these carburetors frequently end up with over-rich mixtures overall because people find out what main jets others are using for similar applications and follow suit. This is not necessarily a good idea, even though the engine might run quite well. An engine that is being fed an over-rich mixture will not deliver maximum power: it might appear to be going very well and you could even be winning competition events with it, but it could be going even quicker with better mixture control.

Holley defines jetting parameters very well, and supplies main jetting that will work on any engine. However, for individual applications there is often some room for reducing main jet size (leaning the mixture off) though usually very little room for increasing main jet size (making the mixture richer).

The main jets need to supply the correct fuel volume for the engine for all part throttle use up to the point when the power valve opens. When the throttle is wide open and maximum power demanded, the power valve will open (because vacuum is sufficient to meet its rating). At this point the fuel supply from the main jet (determined by its size) and additional fuel supplied to the main jet well via the power valve restriction channel (PVRC) drilling in the metering block will, together, provide the optimum full power mixture to allow maximum torque through to maximum rpm. Holley calibrates the carburetors to suit expected amount of air flow (measured in CFM) and this is why sizing the carburetor correctly in the first place is so vital. The closer the CFM requirements are of the particular engine to the CFM rating of the carburetor, the less likely there are to be any major tuning problems.

The biggest problem in determining optimum main jet size is how do you know if the fuel supply is too rich or too weak at full throttle (main jets and PVRCs supplying fuel), and at all intermediate throttle positions when the main jets alone are supplying fuel?

Before embarking on a testing programme to check the fuelling, it is essential to understand that an engine uses the largest amount of fuel per revolution at maximum torque which, on most standard/stock engines, occurs at about 2200-2700rpm. Maximum torque is developed at the point of maximum 'charge density' with the throttle wide open and the engine under maximum loading. On engines that have been modified for higher performance, maximum torque occurs at higher revs (anything from 3000-5000rpm, on average) and the value of the torque will frequently increase as a percentage of rpm. Increases of between 5% and 20% are possible through conventional performance tuning modifications. This means that getting the air/fuel ratio right in the crucial rpm range is critical if the engine is to develop maximum possible torque. After this point is reached, the charge density slowly reduces as the rpm rises and the value of the torque

TUNING 2300, 4160 & 4150 CARBS

reduces by a corresponding amount.

It is through the rpm range where maximum torque is being generated that the degree of ignition advance must be absolutely correct (too much and the engine will 'ping' ('pink') and torque will be lost; too little and optimum torque will not be produced). To all intents and purposes maximum torque requires the use of the maximum amount of ignition advance that the engine will ever need. The optimum amount of ignition advance is that amount, for any given engine design, that causes it to produce maximum torque. However, increased ignition advance seldom has any major effect on a highly efficient design of engine because of the factor of the 'charge density' reducing very slowly from the point of maximum torque as revs increase.

Engines vary in their ideal air/fuel ratios for maximum torque through to maximum rpm full power. Consider 12.2:1-12.5:1 air/fuel ratio, or Lambda 0.82-0.85 or 6.3%-5.0% CO as being the commonly accepted maximum for carburetor-equipped/naturally-aspirated engines.

Another aspect of engine performance is that after the point of maximum torque has been reached, brake horsepower carries on increasing. The reason the brake horsepower increases is because the number of power strokes per minute is increasing. As a consequence, brake horsepower is low at low rpm and high at high rpm. The cylinder filling (volumetric efficiency) performance is reduced after the point of maximum torque, but reduces slowly. As revs rise, there comes a point where volumetric efficiency starts to reduce dramatically; brake horsepower peaks and then, even if the revs go higher, brake horsepower starts reducing. This happens because there simply is not enough time to fill the cylinders enough to cause the brake horsepower to increase. The point of maximum brake horsepower for most pushrod operated overhead valve V8 engines, for example, is almost always at 7200rpm, with a few exceptions going to about 7400rpm). Throughout the rev range and right up to the point of maximum power, fuelling must be correct.

Initial road or track testing to establish basic main jetting requirements

A combination of main jets and the power valve both passing fuel into the main jet well is what gives the engine its calibrated fuel supply for full power. The part throttle mixture (power valve closed) is supplied by the main jets alone. These two aspects of the fuelling must be optimum if the engine is to run perfectly throughout the full rpm range and under all load conditions. The power valve fuel supply is fixed via the size of the power valve restriction channels in the metering block. Therefore, the adjustable variable for altering the overall full power fuel supply is, strictly speaking, the size of the main jets.

If you don't want the expense of rolling road dyno testing, you can test the suitability of various main jet sizes simply by timing how long it takes to accelerate the car to a particular speed. All that is needed equipment-wise is a stopwatch and the car's own speedometer. Note that the engine should be at normal operating temperature at the beginning of every test run.

If under (untimed) full throttle acceleration the engine hesitates, feels reluctant to accelerate or backfires (always audible) through the carburetor these symptoms are an indication of a lean mixture. **Caution!** Don't persist with the test: fit larger main jets immediately. Jet up in stages of two main jet numbers at a time until the backfiring through the carburetor stops.

With the engine accelerating well with no hesitation or backfiring through the carburetor, time how long it takes to accelerate the car to your chosen fixed speed, using the car's own speedo to indicate when the speed is reached. Now decrease the main jet size by two numbers and try the test run again, noting the time taken to accelerate the car to the same speed, and whether there is any hesitation, reluctance or backfiring through the carburetor. If all is well, decrease the main jet size by another two numbers and test the acceleration again. What you are looking for is the fastest performance on the stopwatch, accompanied by urgent, but smooth acceleration without hesitation.

If an engine appears to run equally well on two sizes of main jets, for example 69s and 71s, the final choice must be dictated by part throttle engine response and performance (before the power valve opens). With the engine operating on part throttle, the engine must not hesitate or misfire at any stage before the power valve opens, which will be when the accelerator pedal is getting near the floor. Sometimes, it's necessary to run main jets which give optimum part throttle engine response, but cause the engine to run slightly rich at the top end. This is acceptable if the amount of richness is not noticeably affecting the wide open throttle engine performance.

If you do experience the problem of over-richness at full throttle allied to optimum part throttle performance, an option is to fit a metering block from another model of carburetor which has slightly smaller power valve channel

SPEEDPRO SERIES

restriction holes (0.005in/1.25mm to, perhaps, 0.010in/0.25mm smaller but no more), it just depends on how much richness you are getting. There is some risk attached to this, in that the idle jetting and progression aspect of the metering block might not be as good as the one you remove from the carburetor. Use any primary metering block that proves to be in standard condition, but which has slightly smaller power valve channel restriction holes. Chances are that the idle circuitry, and so on, will be sufficiently similar not to cause undesirable running problems. Alternatively, you could send your original metering block to a Holley carburetor specialist who makes replaceable jets for the power valve channel restriction holes. The carburetor specialist will machine your metering block to take its replaceable jets and send it back to you with a suitable range of smaller jets. Experimentation with combinations of main jets and, if necessary, power valve channel restriction hole sizes, will result in the optimum fuel supply curve being established. You'll now be able to negate the massive richness that you can sometimes get when the power valve opens.

Expect this checking and testing to take some time, though the results will be rewarding. Going to these extreme lengths should not be necessary for most general applications.

If you want to double check any road or track testing work, take the car to a rolling road operator and find out what the air/fuel mixture ratio, Lambda or CO percentage readings are from idle through to maximum rpm/maximum power. Expect the idle mixture to be anything from Lambda 1.00-0.88 or 14.7-13.1 air/fuel, or 1.0-3.5% CO; the full power/maximum rpm mixture readings to be within the range of 0.82-0.85 Lambda or 12.2-12.5 air/fuel ratio, or 5.0-6.3% CO; and the cruise mixture/partial throttle readings to vary between 0.95-1.05 Lambda or 14.0-15.5 air/fuel or 1.5-0.0% CO. A check that covers the mixture strength in this manner will ensure that the full power mixture is correct with no possibility of engine damage through using an over-lean fuel mixture under full throttle acceleration.

Note that, just because an engine is being used for racing purposes only, that does not mean that you can/should make the engine run overly rich. This will only make the engine less than optimum in response, and will use an excessive amount of fuel. Consider the maximum idle speed mixture you would ever want to use on a racing engine to be about 0.88 Lambda - 13.5 air/fuel - 3.5% CO.

Rolling roads are excellent devices for checking the air/fuel ratio of your engine, in conjunction with some form of scientific analysis. No two rolling roads ever seem to read the same, of course, and not all scientific diagnostic equipment is similarly calibrated, but this doesn't matter. What does matter is that at the point of maximum power, the air/fuel, or Lambda, or percentage CO reading is noted, together with the point at which the maximum power drops off. The point of maximum power can clearly be seen on the rolling road's dial, and the maximum power versus the air/fuel mixture ratio required to obtain the given amount of maximum power is unmistakable. Keep going richer until the maximum power starts to reduce, then reduce the main jetting to the next size down and check the figures again to make sure that you can duplicate the maximum power reading. This method will give you the optimum top end jetting. Note that, with Holley carburetors, 0.001-0.002in in main jet diameter will make a difference.

If you make changes to your engine's jetting, you are strongly advised to go back to the same rolling road operator for testing (for reasons of consistency). Determining air/fuel mixtures isn't rocket science, of course, but the biggest problem is knowing what to aim for on the basis of mixture requirements from idle through to maximum power.

The usual methods of checking carburetor tuning are air/fuel ratio, Lambda, and CO emissions levels. Air/fuel ratio meters and CO levels were used for years until the advent of two-way catalytic converters designed to deal with CO and HCs. Another method clearly had to be found, as the COs being registered were no longer necessarily true values. The method chosen involves measuring the oxygen content in the exhaust gases (Lambda) which is indicated on a voltage meter.

It doesn't really matter what scientific diagnosis regime is used by an engine tuning concern. All that matters is that the engine is tuned correctly and optimum power values obtained.

Mixture strength values are often used for comparison purposes. For example, when a power run is completed, an engine might register 390bhp with a 6.0% CO exhaust gas analysed mixture strength. With a richer mixture, it might register 398bhp with a 6.3% CO exhaust gas analysed mixture strength. Taken further, the same engine might register 392bhp with a 6.6% CO exhaust gas analysed mixture strength. Clearly, the 6.3% CO setting was the optimum one. You don't necessarily need to know the CO reading, but it helps to determine the direction you are going in. To this end the accompanying chart with close approximation values has been correlated so that direct comparisons can be made.

TUNING 2300, 4160 & 4150 CARBS

Caution! You must keep an eye on the temperature of the engine when rolling road dyno tuning, as things can get out of hand. The ideal temperature for doing a power run is 75-80 degrees C/167-175 degrees F. If the temperature gets up to 100-105 degrees C/212-217 degrees F, the engine is too hot. The maximum power will likely drop off by 5-10% at the higher temperatures. A consistent and optimum testing temperature is a requirement. Big fans with ducting which can move a lot of air are universally used for rolling road dyno testing.

In the end, however, there is no substitute for road testing. It is possible to come away from a rolling road establishment with an engine that is not going as well as it might (slightly too rich). The ultimate test of engine performance is a road or track test with the power delivery being 'felt' by the driver.

POWER VALVE SELECTION

The power valve is, in effect, a vacuum-operated switch, the rating of which can be used to change the point when extra fuel is going to be supplied by the power valve system. Substituting a 10.5Hg power valve for the standard 5Hg one on a 500-CFM equipped engine, for example, will open the power valve circuit earlier, while fitting a 2.5Hg one will cause the power valve circuit to open later - meaning that the main jets will be supplying all of the fuel until the throttle is almost wide open (mixture could become slightly lean until the power valve opens). Experimentation with power valves will find the optimum rated unit irrespective of the capacity or the degree of modification of the engine.

The power valve that allows the engine to have the maximum possible tractability is the one to fit. Note that for racing applications, there's often little reward for making power valve changes because the 'pedal to the metal' nature of racing meaning that the main jets and the power valve channel restriction holes are both flowing fuel and supplying the ideal air/fuel ratio to obtain maximum power.

It is unlikely that a power valve rated higher than 9.5Hg will be needed on a road going engine. For example, if you have a 10.5Hg power valve fitted, it will open the instant the Hg drops below this figure, perhaps causing the engine to be heavier on fuel than it needs to be and for no real gain. What happens is that the power valve opens, extra fuel is sucked into the engine but it's actually a little too much and so some gets exhausted as partially burnt fuel. Black smoke from the exhaust pipes is evidence of an over-rich mixture. Experimentation is required to find the right power valve to suit the individual application. However, very frequently the factory specified power valve for the carburetor List Number will prove to be the right one!

For the most efficient combustion the power valve should have the lowest rating number conducive to good all round engine performance (meaning that the engine runs on the mains jets alone as much as possible). It is, of course, no use at all running the engine lean anywhere in the rpm range. You'll know when you have gone too far in the lean direction as the engine will momentarily hesitate at the point when the main jets are no longer able to supply enough fuel for the engine but the power valve circuit has not yet come into operation. When the power valve circuit comes into operation the hesitation stops. Changing the power valve for a higher numbered one will cause the power valve circuit to come into operation

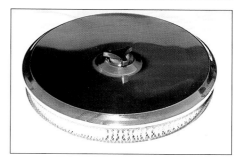

Typical larger diameter aftermarket air cleaner system.

earlier and almost always prevent this temporary problem happening.

Important! Do not confuse retarded idle speed ignition timing, wrong total ignition timing settings, incorrect ignition rates of advance, too low an idle speed setting or long duration camshaft 'fussiness' with carburetor problems. It's all too easy to blame the carburetor when, in fact, it's working fine.

One criteria for choosing a power valve is to do with how much vacuum the engine generates at its usual idle speed. A standard engine will commonly produce about 12 to 14Hg of manifold vacuum at idle, while a medium modified engine (e.g. one fitted with a 285 degree duration camshaft), will produce about 9 to 10Hg. Full race engines (e.g. with 300 degree, or more, duration camshafts), will often produce about 7 to 9Hg of inlet manifold vacuum. The amount of vacuum produced needs to be accurately measured with a vacuum gauge.

Most engines are going to produce between 8 and 14Hg at a sensible idle speed. This means that on an engine fitted with a 6.5Hg power valve, if the throttle is opened a reasonable amount (say, 3/4 open), and the loading on the engine is not all that great, manifold vacuum is not necessarily going drop below 6.5Hg This means the power

SPEEDPRO SERIES

valve will not open and the main jets only will supply all fuel to the engine. If the accelerator is depressed to such an extent that the vacuum does drop below the rating of the power valve, it will open and fuel from the fuel bowl will flow through the power valve, into the metering block passageways, into the main jet well and, therefore, enrich the overall air/fuel mixture supplied to the engine. A considerable amount of extra fuel will flow into an engine when the power valve is open.

At small throttle opening, such as when cruising, the power valve will remain firmly shut. If the engine produces 12Hg of vacuum at its regular idle speed, and between 7 and 11Hg when generally cruising along, a 6.5Hg power valve, for example, will remain firmly shut during most driving conditions. If on the other hand a 9.5Hg power valve is fitted to the same engine and used under the very same circumstances, the chances are that the engine will use more fuel (10 to 20%) to provide the same performance. This is because when the manifold vacuum drops below 9.5Hg the power valve is going to open and enrich the overall air/fuel mixture. The power valve could well be opening up and closing very frequently and running a very rich overall mixture for no appreciable power gain. This is not the way to good economy and it would be better to fit a lower numbered power valve to make sure that it stays firmly shut until acceleration power is needed.

For road going cars to get the best level of economy, the engine needs to run on the main jets only as much as possible. To achieve this the main jets need to be just large enough to run the engine under all normal part throttle circumstances, and the rating of the power valve low enough to stay shut as much as possible so that it only opens when power is really demanded.

It takes some experimenting to achieve this but it can be done, and very good fuel economy is possible.

Why engines fitted with long duration camshafts need low rated power valves

Engines fitted with very long duration camshafts (295-300 degrees plus) are not noted for producing a lot of inlet manifold vacuum at low engine speeds, and what they do generate will fluctuate considerably. What this means is that if an engine is fitted with a 6.5Hg power valve and the manifold vacuum fluctuates quite a lot and drops down to, say, 5.5Hg at times, the power valve will open whenever the vacuum drops below 6.5Hg and stay open for as long as the vacuum remains less than 6.5Hg. In this situation extra fuel will be drawn into the engine, the mixture will enrich and the idle will get rough. The amount of extra fuel in the idle mixture could be 50% or more than it should be, which is enough to cause the engine to stall.

The solution to this situation is to find out what the minimum amount of vacuum being generated at maximum engine speed is and fit a power valve which has a lower rating by a minimum of 0.5Hg. Also see if the engine's idle speed can be increased slightly so that the idle is smoother and, therefore, less likely to fluctuate, which will also cause the inlet manifold vacuum to be higher. In most instances these days long duration camshafts are not used so this problem doesn't usually arise.

In conclusion, when very long duration camshafts are used, the power valve rating should be below the minimum amount of vacuum the engine generates at idle. While fitting a power valve with an appropriate lower rating will prevent the power valve opening and closing erratically and, possibly, causing idle speed richness, ultimately it may not prove to be the best rating for when the engine is operating in its working range. In this rare situation it is essential that the right power valve for all engine speeds above idle is used, even if idle quality has to be compromised.

FUEL BOWL LEAKS

This is a common, but unnecessary, problem with Holley carburetors which you really do need to be prepared for. The two common places where fuel bowls leak are from the two bottom securing screws and from the gasket between the fuel bowl and the metering block. Most fuel leaks are caused by using 'one time' gaskets more than once, and reusing washers which are not serviceable. **Warning!** Fuel bowl leaks are potentially very dangerous: take **no** risks with fuel.

The fuel bowl to metering block gaskets are very vulnerable to damage, but seldom leak if they're renewed every time the fuel bowl is removed and replaced. You may be tempted to see if the gasket will go another time. They normally don't.

The two lower securing screws have fibre washers between the undersides of their heads and the fuel bowl. These washers will last for several bowl removals and refitting but, in the end, will fail and replacement is the only option. Most people use them until they first show signs of needing too much tension to be applied to the screws to stop fuel leaking out. However, all engines should have new washers fitted after testing and jet changes to avoid any possibility of fuel leakage. The two top securing screws are less critical with regard to fuel leakage, but the washers still need to be in good condition. It is essential to have spare float bowl washers on hand as well as a fuel bowl to metering block gasket.

TUNING 2300, 4160 & 4150 CARBS

SUMMARY

1 - Always check that when the throttle pedal is fully depressed the carburetor butterflies are vertical. This is one of the most common faults causing a lack of power and acceleration (engine not getting full throttle). Check this factor frequently and certainly on race day before racing starts.

2 - Set the idle speed at a suitable level for the engine (particularly if it has been modified): not too slow and not too fast. Consider 500 to 700rpm to be suitable for a standard engine, 700 to 900rpm for a c.270 degree smooth idle camshaft, and 1000 to 1200rpm for a longer duration camshaft. Some very long duration camshafts (lots of valve overlap) will not allow the engine to idle at less than 1500rpm.

3 - Check the fuel pressure and make sure that it is able to be maintained by the pumping system, especially on a long straight on a race track at maximum rpm with wide open throttle. **Caution!** Fuel starvation can cause a dangerously weak mixture.

4 - Check the ignition advance setting of the engine frequently, and certainly on or before a race day. Retarded ignition is a major cause of poor engine performance. Consider most V8 engines on average to need 12, 14 or 16 degrees of idle speed advance, and 34, 36 or 38 degrees of total advance.

5 - Make sure that the fuel filter is clean and not clogged. A blocked fuel filter can cause a reduction in the fuel level. This is why the take off to a fuel pressure gauge should always be after an in-line fuel filter. **Caution!** Fuel starvation can cause a dangerously weak mixture.

6 - Does your carburetor take in cool or heated air from the engine compartment? See what can be done to ensure that cool air circulates around the carburetor and the engine compartment. Can ambient temperature air be fed to and around the air filter? Route the fuel pipes clear of any heat source. Insulate the carburetor from engine heat as much as possible.

7 - Have a range of carburetor gaskets on hand, just in case there is a failure. Chasing around at a race meeting looking for a replacement gasket is a complete waste of time and races have been missed for the sake of a gasket.

8 - **Warning!** Always carry a suitably sized fire extinguisher appropriate for fuel fires in the car, and have it near to hand whenever you are working on the carburetor. Many a car has been lost through not being able to deal with an insignificant fire which has grown out of control. Gasoline is dangerous and must be treated with caution at all times.

9 - Fit at least two good return springs to the throttle arm of the carburetor. On a V8 engine, for example, one spring needs to be mounted on or near the front of the rocker cover and go back to the top of the throttle arm, while the other spring needs to be mounted on or near the back of the same rocker cover and go to the bottom of the throttle arm. Both springs are trying to shut the throttle. The tension of the springs must be such that they both apply good closing tension without making the throttle pedal unduly difficult to push to the floor. Both springs will need to be in the region of 6 to 7 inches long and be of equal tension.

10 - Always use a large paper air cleaner system on your engine. Large aftermarket air cleaners are available for four barrel and two barrel carburetors. If the seal between the paper element and the pressed steel base and top seems doubtful to you, use silicone sealer on the paper element faces that contact the pressed steel top and bottom sections of the air cleaner body. **Caution!** Keeping dirt out of your engine is paramount to maintaining maximum engine life and good performance. Large paper air filters are very efficient.

11 - Use a manual choke if a choke is required. Removing the choke plate from the air entry of the carburetor is not going to make any appreciable difference to the engine performance in most cases and certainly not on a daily driven road car.

12 - Fit new gasoline rated fuel hosing and route the pipes well clear of the exhaust pipes. Fit close fitting stainless steel worm drive hose clamps to all hose connections. Consider routing the fuel pipe through a thick walled steel pipe if it passes close to and in line with the clutch bellhousing and is otherwise unprotected. It is not impossible for a shattered clutch to explode through the bellhousing and sever a fuel line.

FURTHER TUNING OF FOUR BARREL CARBURETORS

With the primary barrels of four barrel carburetors correctly tuned, it's time to unlock the secondary barrel throttle mechanism and continue the tuning procedure as detailed in chapter 6.

SPECIFIC INFORMATION ON 2300 TWO BARREL HOLLEY CARBURETORS

In general these two barrel carburetors are very underrated, mainly because they simply don't look as good on a V8 engine as a four barrel Holley carburetor. The fact that they can often give equal engine performance in some applications just doesn't come into it. In fact, many smaller V8s will go amazingly well with a 350 or 500CFM two barrel carburetor used in conjunction with a good inlet manifold. Large capacity four cylinder and six cylinder engines can also use these carburetors with good results.

SPEEDPRO SERIES

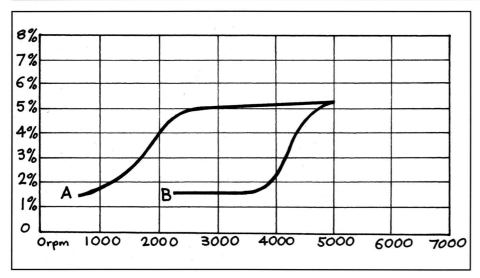

A standard production type pushrod V8 engine: line A is a basic fuel curve from idle to maximum power; line B indicates the cruise fuel curve (minumum accelerator pedal) up to where the accelerator pedal is fully depressed (where the lines intersect).

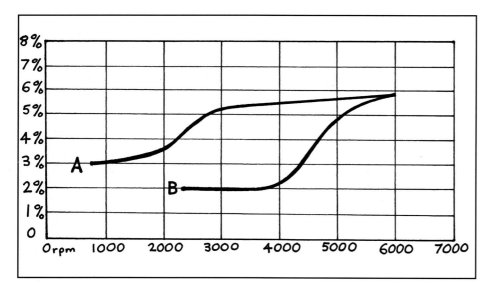

A moderately modified production type pushrod V8 engine: line A is a basic fuel curve from idle to maximum power; line B starts indicates the cruise fuel curve (minumum accelerator pedal) up where the accelerator pedal is fully depressed (where the lines intersect).

Inlet manifolds must be well designed for good breathing. Frequently the original equipment cast iron 180 degree inlet manifold on a V8 engine reduces potential carburetor performance because they were designed to produce maximum torque at low rpm and the engines were expected to operate in a low overall rpm range (500rpm to 4000rpm approx). With a 2300 Holley carburetor fitted on to a stock/ standard type inlet manifold, the improvement in engine performance is often disappointing. In the quest for better performance, off comes both the two barrel Holley carburetor and the original equipment inlet manifold, and on goes a four barrel Holley carburetor fitted on top of an aftermarket aluminum inlet manifold. The improvement in engine performance is noticeable and the two barrel carburetor is often thought of as being part of the earlier lack of performance problem. It seldom is: in fact, the inlet manifold is almost always the big problem. Try the two barrel carburetor with a high rise type 360 degree inlet manifold in conjunction with an adapter plate before going to a four barrel carburetor, you might be surprised ...

If an adapter is to be made out of aluminum plate to mount the two barrel carburetor onto an aftermarket inlet manifold, it frequently pays to fit the two barrel carburetor so that its two front stud holes use the two front studs of the four barrel inlet manifold. This means that the two barrels will be in the same place as the front two barrels of a four barrel carburetor would have been. There are several advantages in doing this. The first is that the large diameter air cleaner that would normally be fitted onto a four barrel carburetor can be fitted onto the two barrel carburetor and still be centrally situated as per normal and not cause any clearance problems. The second is that the throttle linkage will be the same for a two barrel or a four barrel carburetor with no alterations being required if a four barrel carburetor happens to be fitted at a later stage. The third is that V8 engines frequently go slightly better with a two barrel Holley carburetor in this position.

When original equipment carburetors get worn out they're

TUNING 2300, 4160 & 4150 CARBS

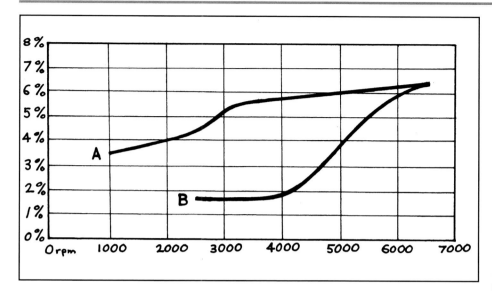

A race prepared production type pushrod V8 engine: line A is a basic fuel curve from idle to maximum power; line B indicates the cruise fuel curve (minumum accelerator pedal) up to where the accelerator pedal is fully depressed (where the lines intersect).

Lambda	Air/Fuel	% CO
0.80	11.8	8.0
0.81	11.9	7.3
0.82	12.0	6.5
0.83	12.2	5.9
0.84	12.4	5.4
0.85	12.5	5.0
0.86	12.6	4.85
0.87	12.8	4.35
0.88	13.0	3.8
0.90	13.2	3.3
0.91	13.4	2.85
0.92	13.5	2.6
0.93	13.7	2.15
0.94	13.8	1.9
0.95	14.0	1.6
0.96	14.1	1.4
0.97	14.3	1.0
0.98	14.4	0.8
0.99	14.6	0.6
1.00	14.7	0.5
1.01	14.8	0.6
1.02	15.0	0.3
1.03	15.1	0.15
1.04	15.2	0.2
1.05	15.4	0.15

Lambda/air-fuel/% CO conversion chart.

often replaced with a suitable Holley carburetor, especially if the Holley carburetor simply bolts on to the manifold. However, don't expect to be able to bolt on a Holley carburetor and to immediately have the engine's performance improved by a huge amount: it just isn't going to happen. Expect an improvement by all means, the Holley carburetor will likely be a better carburetor than the original, and it won't be as worn as the original. If the original carburetor was in a good serviceable condition and well tuned, the chances are that the improvement will be quite small.

For normal road going work, most mass production V8 engines of between 215 and 302 cubic inches will go best with a 350-CFM carburettor. However, if V8 engines in this size range are modified with improved cylinder heads (or better factory heads), 280 to 285 degree high performance camshafts, a four barrel inlet manifold and a good exhaust system, then the 500-CFM Holley carburetor will definitely prove to be better than the 350-CFM because, in this situation the airflow capacity of the 350C-FM carburetor could be exceeded by the engine's requirements. However, up to the point of maximum airflow there will be virtually no difference in engine performance, with slightly better engine response at low rpm possible with the smaller carburetor. Certainly in a competition environment the 500-CFM carburetor is the best option for engines in this size range because they'll be turning 3500 to 6500 or even 7000rpm for much of the time.

It's difficult to resist the idea that the bigger the carburetor the better, irrespective of the circumstances. However, this is definitely not true as is proved in racing when the rules insist on a certain size of carburetor: 500-CFM carburetors have been used on surprisingly large engines turning quite high rpm with quite remarkable results.

Some V8 engine manifolds will allow these Holley carburetors to bolt straight on. Small block Fords are good examples of this. The bolt pattern of the Holley two barrel is the same as the FoMoCo two barrel carburetor. All things being equal, the FoMoCo carburetor equipped engine will not perform as well as the 500-CFM Holley equipped engine even though the internal workings of both carburetors are very similar. The FoMoCo carburetor will, however, usually return slightly better economy than the Holley.

Be aware of the fact that many (but not all) standard mass production inlet manifolds do not flow air/fuel mixture all that well in a high-performance and a high rpm application, and a significant

SPEEDPRO SERIES

amount of engine efficiency can be lost through continuing to use them. Such manifolds can cripple the carburetor's performance potential. A Holley carburetor can be bolted onto a poor standard inlet manifold and give no appreciable gain in power, yet this is not necessarily the fault of the carburetor. Fit that same Holley carburetor onto a good aftermarket inlet manifold (or a better type of factory inlet manifold) and you'll see engine performance increase (by up to 10%). Good inlet manifolds are vital if maximum efficiency (especially acceleration) is to be obtained from a Holley carburetor equipped engine.

If you're going to use a two barrel Holley you need to be aware of the importance of good manifolds and select a good 'high rise' 180 or 360 degree four barrel inlet manifold with reasonably sized inlet manifold runners. Not huge runners, but a little larger than those of the standard manifold you're replacing. Some inlet manifolds of the same basic configuration and design are better than others and this is because of internal runner sizing and smoothness of contour. Most high performance inlet manifolds (aftermarket or factory optional) are very good.

Holley, and other companies, makes adapters that bolt on to four barrel inlet manifolds so that a two barrel carburetor can be fitted easily on any V8 four barrel inlet manifold. Holly's own adaptor is part number 17-10 or 17-12 (depending on the original bolt pattern of the inlet manifold).

To illustrate what has already been written, when fitting a 500-CFM Holley to a small block Ford V8 engine, change the standard cast iron inlet manifold for a 360 single plane one, such as the Holley 3006, Edelbrock Torker 289, or similar, aftermarket inlet

'Power valve channel restrictions' (PVCR) being checked with the shank of a drill bit to make sure that the sizes are correct in this metering block.

manifold. Use a Holley adapter plate or make up your own. The difference in engine performance will be quite noticeable: even more noticeable if the engine is a reasonably well modified one. Any small block Ford engine, for instance, which has modified cylinder heads, a 280 to 285 degree duration camshaft, a good exhaust system and a correctly set-up ignition system is going to pull strongly to 6500-6700rpm. The same engine fitted with a FoMoCo carburetor does not go anything like as well. The ability of Holley carburetors to give better engine performance than most other similar carburetors is not really in doubt, provided the Holley carburetor is set-up correctly. However, the Holley equipped engine will use more fuel when power is demanded.

Note that the starting point jetting to have in either of these two barrel carburetors is the standard Holley recommended jetting for the List Number. Changes are only made after thorough testing and proof that

the jetting is clearly not correct for the engine and application. The standard Holley jetting is in this instance as follows:

Holley 350-CFM Universal Performance two barrel carburetor List Number 7448: comes from Holley with a centre pivot float bowl, a manual choke, 62 main jets, stock capacity 30cc accelerator pump, a white 248 accelerator pump cam, a 31 accelerator pump discharge nozzle and a 8.5Hg power valve. The 7448 is the best version of the 350CFM to buy because, besides being road ready, it is race ready as it comes.

Holley 500-CFM Universal Performance two barrel carburetor List Number 4412: comes from Holley with a centre pivot float bowl, a manual choke, 73 main jets, high capacity 50cc accelerator pump, brown 336 accelerator pump cam, a 28 accelerator pump discharge nozzle and a 5.0Hg power valve. The 4412 is the best all round version of the 500-CFM to buy because besides being road

TUNING 2300, 4160 & 4150 CARBS

The original power valve channel restriction hole has been drilled out and tapped. The small jet that screws in is arrowed.

The two jets screwed into this metering block (arrowed). The tops of the jets must be flush with the surface, otherwise the power valve will not seat correctly.

ready, it is race ready as it comes.

There was also a 650-CFM model of these carburetors made for a few years under List Number 6425. While rare, they do occasionally turn up at swap meets and in small ads.

All replacement spare parts are still available. These larger CFM rated Holley carburetors will offer improved engine performance, but only if the engine really can use the extra flow capacity. Check out the Holley calculation for engine size in relation to CFM before buying a 650-CFM carburetor.

Accelerator pumps

The 500-CFM two barrel carburetor comes as standard with a 50cc accelerator pump diaphragm. This is seldom necessary for road going engines, and the 30cc accelerator pump is usually quite adequate. There is nothing wrong with retaining the 50cc accelerator pump, it's just that it is rarely necessary or desirable to have slightly too much fuel going into the engine for optimum acceleration phase performance. If all is well during testing, keep it, if not change it for the smaller unit. On the basis of sharpness of response and fuel economy, fitting the 30cc pump diaphragm can be quite beneficial.

Caution! In many instances the 50cc accelerator pump cover and actuating lever may foul the inlet manifold. In some cases the carburetor will have to be packed up with a metal spacer, or two or three gaskets or more, to clear the inlet manifold. Alternatively, the inlet manifold can be relieved by grinding material away so that the cover and actuating lever have adequate working clearance. Always check this feature when fitting a carburetor onto any inlet manifold.

In the first instance, for a 500-CFM carburetor, set the accelerator pump up using the standard 50cc diaphragm and lever assembly. However, be aware of the possibility that the bigger pump shot could be the cause of acceleration phase problems if it is too long in duration, making a

30cc accelerator pump a better option. There's not usually any advantage in swapping the standard 50cc accelerator pump for a 30cc one on engines with capacities of 302 cubic inches (5 litres) or more.

Main jets

The ideal main jet sizes for these two carburetors are, on average, always going to fall between 59 and 73 when fitted to the usual range of engine capacities that these two 2300 carburetors are capable of fuelling (2.5 to, perhaps, 6.0 litres).

Power valves shutting at high rpm

Caution! A possible problem that can cause engine damage is to do with the prospect of the power valve closing and cutting off the supply of extra fuel at high rpm. This can happen in classes of racing where the carburetor is size (CFM capability) limited or when the carburetor is too small for the particular engine.

The reason that the power valve closes with the throttle wide open is that the inlet manifold develops a vacuum because the carburetor is restricting airflow. If the amount of vacuum generated by the engine becomes more than the rated amount of the power valve, the valve will definitely close with possibly disastrous results (engine damage through mixture weakness). To avoid this situation check, using a vacuum gauge, whether a vacuum is present in the manifold at wide open throttle and the maximum rpm you are ever likely to use. If necessary, fit a power valve with a higher rating number. **Caution!** The power valve must never be allowed to close at full throttle and maximum rpm.

It is quite possible to fit a power valve plug and not use the power valve circuitry at all, using only the main jets for maximum power. Obviously the main jets must be larger to compensate for the lack of additional fuel that would normally be supplied by the power valve.

Engines set-up to run without a power valve are frequently over rich at low revs as the air fuel mixture ratio has to be set for top end running using the mains only. Of course, in some applications this just doesn't matter because the engines are jetted for top end power alone. However, most engines will definitely run better with a power valve fitted. Holley recommends that primary power valves be fitted in all circumstances. The reasoning is that while power valves can be removed, a better fuel curve is possible with the power valve fitted.

Irrespective of which power valve is fitted to a Holley carburetor, when the throttle is fully depressed there should be zero vacuum being produced in the manifold, and the power valve should be open to provide maximum fuelling.

FOUR BARREL TO TWO BARREL OPTION

While many may frown at what follows, the fact is that the system described will work very well. In many instances a secondhand four barrel carburetor is available for a very reasonable price, or you may already have one. Instead of using a two barrel Holley carburetor consider using half a four barrel carburetor (primary side only). If the engine you are using is a reasonably small V8, such as a 3.5 to 5.0 litre, a 750 to 780-CFM four barrel Holley will almost certainly be too large for the engine, so using half of it is a possibility.

Using the butterfly and venturi sizes of the various large four barrel carburetors for comparison to the 500-CFM two barrel carburetor (divide a four barrel's CFM rating by 1.5),

Always make sure that the fibre washers as fitted to the two lower fuel bowl securing screws are in excellent condition (arrowed). Have on hand spare ones in case they start to leak.

TUNING 2300, 4160 & 4150 CARBS

it's clear that the primary barrels of a 750-CFM four barrel carburetor are equivalent to a 500-CFM two barrel carburetor. There are several other CFM ratings available if the full range of the large four barrel carburetors is taken into consideration.

There are some advantages in using a four barrel carburetor as a two barrel carburetor unit on a small capacity engine. The four barrel carburetor simply bolts straight onto any Holley carburetor bolt pattern aluminum factory or aftermarket inlet manifold, so no adapter plate system is necessary. Also, large four barrel carburetors (750 - 780-CFM) are frequently available secondhand in perfect condition and at unbelievably low prices.

Warning! - If you do use a four barrel carburetor as a two barrel unit, it's imperative that the secondary barrels are unable to admit any air to the inlet manifold which could cause the air/fuel mix to become too weak. The most secure way to decommission the secondary barrels is to make a sandwich plate that matches the 'footprint' of the carburetor base, except that the plate will contain no openings for the secondary barrels. The plate can be made from 16 gauge aluminum sheeting and be sandwiched between two standard gaskets. It takes a bit of work to make this plate accurately using a gasket as a template, but it only has to be done once. If a mechanical carburetor is used, the secondary throttle linkage must be disconnected/removed. If a vacuum secondary carburetor is used the vacuum hole which connects the primary barrels to the diaphragm must be plugged somewhere. There should also be no fuel supply to the secondary fuel bowls. Nothing about the four barrel Holley carburetor used in this instance has to be irretrievably altered to allow it to function as a two barrel carburetor if a supplementary base plate is made so such a carburetor can be returned to four barrel operation in a very short time.

www.velocebooks.com/www.veloce.co.uk
All books in print • New books • Special offers • Newsletter

Chapter 6

Tuning the secondary barrels of 4150 & 4160 carburetors

In the first instance, all four barrel carburetors should be fitted with the Holley recommended secondary jetting as found in the Numerical Listing for the individual carburetor model. If the carburetor or any of its components have an unknown history check all metering plates and main jets for originality and factory specifications before fitting them into the carburetor. You need to know now whether they have been drilled out or otherwise altered.

Once the primary barrels are tuned correctly (refer to chapter 5), and the engine's running as it should, the secondary barrels can be allowed to operate again by removing the binding wire and reinstating mechanical linkages and so on as necessary.

Four barrel carburetors are divided into two basic groups for secondary barrel tuning: 1) vacuum secondary units and 2) mechanical or 'double pumper' units. These two categories in turn have sub-groupings in that some vacuum secondary carburetors have metering blocks (4150 units) and others have metering plates (4160 units), additionally, some secondary metering block versions of these vacuum secondary carburetors have power valves fitted to their secondary metering blocks. All mechanical secondary carburetors have a secondary metering block, but not all of them have power valves or secondary metering blocks with idle adjustment screws (commonly called 'four corner idling'). So, there are three types of secondary metering block used on mechanical secondary carburetors, with the 'four corner idle' type referring to some mechanical four barrel carburetors having the same type of metering block fitted front and rear.

The Holley Numerical Listing does not directly specify which 4150 vacuum secondary or mechanical secondary carburetors have secondary power valves and which do not. The way to work out which carburetors have secondary power valves, and what the Hg values of those power valves are, is to note the figure in brackets adjacent to the primary power valve part number. The figures to look for are (12), (15), (22) and (30) or (both): 12 = 8.5Hg, 15 = 6.5Hg, 22 = 3.5Hg and 30 = 2.5Hg power valves, and '(both)', meaning the secondary is the same as the primary power valve. If there is no reference to power valve figures in brackets, the carburetor on that line doesn't have a secondary power valve fitted to it.

The Numerical Listing also has a column in it titled 'secondary nozzle size or spring color'. This column is applicable to all 4150 and 4160 vacuum secondary or mechanical secondary carburetors. The listing gives the standard Holley discharge nozzle sizes for mechanical carburetors and vacuum secondary diaphragm spring colors for vacuum secondary

TUNING THE SECONDARY BARRELS OF 4150 & 4160 CARBS

carburetors. When a nozzle size is listed, you can assume the carburetor on that line is a mechanical secondary one as vacuum secondary carburetors do not have secondary accelerator pumps. Conversely, when a vacuum secondary diaphragm spring is listed, the carburetor on that line is a vacuum secondary one. It takes a bit of familiarisation to understand the mass of information contained in the Numerical Listing. The amount of information that can be gleaned from it, together with the *Holley Performance Parts Catalog*, is not immediately obvious to a new user.

In the first instance check the float level of the secondary fuel bowl in the same way that the primary fuel bowl was checked (refer to chapter 5 for the details). Like that for the primary fuel bowl, this adjustment is vital for the smooth running of the engine. If the fuel level is too high the mixture will be too rich and if the fuel level is too low the mixture will be too lean, all other factors being equal.

The diaphragm of vacuum secondary carburetors must be checked to see that it is air tight. If there's a leak it won't hold the required vacuum. With the diaphragm assembly off the carburetor, push the diaphragm shaft fully into the housing, and hold it there; then place your finger over the housing's inlet hole. Release the diaphragm shaft: it should not move out from the housing although it may move to one side slightly. If it does move outward from the housing there is leak in the diaphragm that renders the vacuum's secondary barrels unserviceable. Don't proceed with tuning the carburetor until this problem is rectified.

Check the accelerator pump adjustment on mechanical secondary carburetors in the same way that the primary barrel accelerator pump was adjusted. That means no up and down lash (free play) in the accelerator pump lever arm where it contacts the throttle linkage arm, though there must be some sideplay. This is a 'tricky' adjustment that has to be got absolutely right. Failure to get it right results in hesitation during acceleration, even if the cam, cam position and nozzle are the correct ones. Select one of the two cam positions and test the engine. Then change the cam position to the other setting and test the engine again. **Important!** Remember to re-adjust the accelerator pump lever after any cam change or cam position change.

When the secondary barrels are re-activated, make sure that the four butterflies of a mechanical/double pumper carburetor are vertical at full throttle; that the two primary butterflies are vertical on a vacuum secondary carburetor when the accelerator pedal is fully depressed, and that the secondary barrels can be fully opened manually. This involves looking down the venturis with a flashlight (torch), if necessary, to see if all the butterflies are in the vertical position.

Sometimes the stops on the throttle body and shaft assembly are not right and prevent the butterflies opening to vertical. To rectify this problem, it's easier to remove the carburetor from the engine and remove the throttle body and shaft assembly from the body of the carburetor to get at the stops. With everything easy to get at, the stop can either be hand filed, or material removed with a high speed grinder fitted with a rotary file or a mounted point grinding stone. Usually, when the butterflies don't open fully, the throttle linkage is at fault and a simple adjustment will sort the problem out. Butterflies always need to be checked to ensure that they do open fully as much power can be lost if this adjustment is not correct. Repeat the check fairly regularly as throttle linkages seem to have a habit of moving out of adjustment.

USING ACCELERATION TIMING TO OPTIMISE JETTING

A full throttle acceleration test will show up any carburetor problems almost instantly. Testing the engine with the secondary barrels is, essentially, identical to the testing procedure used to set the primary two barrels for optimum performance.

Warning! Acceleration tests are best conducted off the public highway. Never do acceleration tests alone and always have a fire extinguisher in the car: ensure that whoever accompanies you has been instructed in how to use the fire extinguisher.

Warning! Always test with the air cleaner fitted. The air cleaner stops flames from backfires going any further than the top of the carburetor. The large round paper type air cleaners are ideal for use on these carburetors and the bigger in diameter the better. There are three general diameters and two fitted heights of housing. Paper type filter element air cleaners are very efficient, in-expensive and reasonably compact. Although the two metal pressings of these filter housings are almost always quite accurately made, there is still the possibility of dirt and so on getting into the inside of the filter generally. The way to prevent any dirt getting into the carburetor intake area is to smear a generous amount of silicone sealer over both sealing edges of the filter and then fit the air cleaner together and allow the silicone sealer time to dry (15 minutes). Change the filters at regular maintenance intervals. These paper air filters are as good as it gets for air filtration and they're not restrictive. Engine manufacturers use paper as a filtration material because it is virtually unmatched for air cleaning efficiency.

Caution! Before acceleration testing make sure that the top end mixture is correct to avoid any possibility of engine damage resulting from an overly weak mixture. If the stock Holley recommended secondary jetting is fitted to the carburetor, it's most unlikely that the fuel mix will be lean, it just might not be optimum. Don't take anything for granted, check the carburetor jetting in relation to Holley's Numerical List before testing the engine.

For some people the peace of mind of having the overall mid-range and top end fuel mixture checked by scientific means (exhaust gas analysis or air/fuel ratio reading) before a road or track test is important. If you feel this way, the engine should have the wide open throttle fuel/air mix checked on a rolling road dyno. This can ensure that with all four butterflies wide open, the mixture strength is not too rich and not too lean. Also, any basic carburetor fault will be shown up. If the mixture is too lean, for example, this can be disastrous, with engine damage possible; while, if the mixture is too rich, optimum power will not be produced and fuel consumption will be excessive. With a wide open throttle, full power, air/fuel mixture ratio of 12.2-12.5:1/Lamda 0.82-0.85/6.3-5.0% CO, the engine can be road or track tested with the sure knowledge that the fuel/air mix is not too lean at wide open throttle. This does not mean that the jetting is finalised, although it is likely to be. Finalising jetting is **always** determined by thorough road or track test, or by the quickest time taken to cover a set distance.

Another way of testing carburetor efficiency is to see how long it takes to accelerate the car using top gear only from, say, a steady 3000rpm to 6000rpm, or from 3500rpm to 7000rpm, depending on the engine's useable rev range. A lower gear (3rd gear of a 4 speed gearbox, for instance), could also be used, and this will cut the speed down and the test results still end up equally relevant. **Warning!** If high speed is to be involved, a safe venue must be used and all precautions taken. What is being measured is the ability of the engine to accelerate from mid rev range to just under maximum power in one gear (gearchanges could affect the procedure's result). Generally speaking, this testing procedure is a highly accurate measure of a carburetor's/engine's performance.

To establish a baseline acceleration time (since on any particular day atmospheric conditions can affect engine performance), test the engine with the front two barrels of the carburetor only in operation. Repeat the test several times to establish an accurate timing. Reconnect the secondary barrels and repeat the test a couple of times. If the elapsed time increases, something is wrong with the secondaries. In this case, check the rear part of the carburetor for a general fault and reconfirm the recommended Holley jetting is installed by going through the Numerical Listing.

On the first test with the secondary barrels in operation there could be some hesitation, or a flat spot. This will most likely be caused by the diaphragm spring being too light on a vacuum secondary carburetor, or the accelerator cam, cam setting or nozzle size not being quite right or adjusted correctly on a mechanical secondary carburetor. If the hesitation clears as the revs rise and the engine accelerates willingly and strongly the main jetting is nearly right, but not necessarily optimum.

In the first instance, concentrate on getting rid of the hesitation. Note that if the hesitation is really very bad there is probably a blockage (or blockages) somewhere in the secondary part of the carburetor, or the volume and pressure of the fuel supply is inadequate. Check the metering block, carburetor body passages, throttle body, shaft assembly passages and the fuel bowl for blockages. Check that the fuel pump is working properly and that any filters in the fuel line are clean.

Vacuum secondary carburetors: if there is hesitation during the acceleration test, fit the strongest diaphragm spring (black) into the diaphragm housing and repeat the test. This spring ensures the slowest operation of the secondaries. If the hesitation or generally poor engine performance persists with the stronger diaphragm spring installed, increase the main jetting by one or two increments. This can be achieved by changing the main jets for richer ones, drilling the main jet holes in the metering plate to a larger size in increments of one jet size, or changing the metering plate for one with slightly larger main jets. **Caution!** - If the hesitation gets worse after richening the mixture, then **and only then** go leaner. Always go for a richer mixture initially. On power valve equipped vacuum secondary carburetors, the power valve will be operating and allowing extra fuel into the main jet well during wide open throttle full power testing. If the power valve is not opening (i.e. it is faulty), the main jets alone will be supplying fuel and the mixture will be excessively lean.

There is no doubt that power valve equipped vacuum secondary carburetors which are set up well

TUNING THE SECONDARY BARRELS OF 4150 & 4160 CARBS

The latest 570 CFM, vacuum secondary Holley has centre pivot fuel bowls, bright surface finish and a clear plastic non-removable fuel level sight glass system.

can give optimum wide open throttle performance and very 'clean' part throttle performance. This is purely because on partial throttle the power valve will be shut, and the secondaries will be running on the main jets only. The overall mixture will be slightly weaker in this situation and this can make a difference.

Single gear acceleration testing from low rpm on a steady part throttle and then 'flooring' the accelerator pedal so that the engine accelerates under maximum loading through to nearly the end of the power surge, will let you find the diaphragm spring that will allow the smoothest/quickest acceleration over a set distance in the shortest time.

The low rpm starting point depends on the camshaft fitted to the engine, i.e., it must be in its power range but only just. Use the lowest convenient rpm starting point. If your engine is fitted with a 'smooth idle' high performance type camshaft, for example, that 'pulls well' from, say, 2200rpm, use 2500rpm as your starting point. This is nothing more than making sure that your acceleration phase testing is not being negatively affected by the camshaft not being in its optimum working range. If that very same camshaft's power surge has a maximum of 5800rpm, use 5500rpm as the maximum rpm for testing. This will ensure that the power surge hasn't finished, since a slowdown in engine acceleration would negatively affect the test. This example gives us a 3000rpm working range for testing purposes, right in the optimum part of the camshaft's power production range.

Check how long it takes for the engine to accelerate from 2500rpm to 5500rpm in one gear (3rd or 4th gear of a four speed gearbox, for example). If the engine hesitates, 'bogs down', or 'spits back' through the carburetor, for example, the driver will 'feel' this and know that something is wrong. You will also know how long it took to go from 2500rpm to 5500rpm. This baseline time, with the 'problem' present, will give you a starting point for further testing, eliminating any problems, and reducing the elapsed time.

This sort of testing regime should be applied to each individual phase of finding of the correct components to use in your carburetor.

Mechanical secondary carburetors: For the first acceleration tests, place the Numerical Listing recommended accelerator pump cam in the 1st position (generally the leaner position). For the second acceleration test, change the cam position to the 2nd position, re-adjust the accelerator pump arm and repeat the acceleration tests. Hesitation is generally caused by the accelerator pump cam not injecting a sufficient amount of fuel into the engine through the recommended nozzle. This problem can be caused because the lever arm is not adjusted correctly or, less likely, the accelerator pump is too small, the recommended nozzle is too small, or the accelerator pump cam is just too small in terms of 'shot' rate delivery for the engine.

Testing the range of accelerator pump cams as listed in the accelerator pump discharge volume chart (from the richer side of the recommended accelerator pump cam) in the 1st and 2nd positions will generally show an improvement in acceleration. Keep going richer and don't miss a step as you might pass the optimum combination. Use the accelerator pump cam that gives the best all round acceleration without hesitation, but avoid going over rich. Always go through the full range of cams before increasing the nozzle size. If none of the cams proves to be satisfactory then, and only then, go up one increment in nozzle size and then go through the range of cams again, starting with the smallest 'shot' size cam. There is always one cam, one hole position and one nozzle combination that will work correctly. It can take some time to arrive at the best accelerator pump cam and cam position for your application, but once done it will not need to be done again. Note that in some rare instances drilling a discharge nozzle to an in-between size is the only way to get the accelerator pump 'shot' absolutely right. However, in most instances, using the factory range of discharge nozzles gives a good result. Setting up the secondary accelerator pump system is slightly less critical than the primary accelerator pump system because the engine is already accelerating via the primary barrels. The settings do still have to be as near to optimum as possible though.

With the stock recommended main jetting fitted, and the carburetor correctly sized for the engine, no Holley carburetor is going to give a grossly lean fuel/air mix. This is why the diaphragm springing can be sorted out before the main jetting is optimised on a vacuum secondary carburetor, and the accelerator pump mechanism setting finalised before the main jetting is optimised on a mechanical secondary carburetor.

Once hesitation free acceleration and the quickest elapsed time is achieved, the next step is to increase the main jet sizes in the secondaries and then to repeat the tests. If the times reduce, keep on increasing the main jet sizes by 2 numerical increments at a time on main jets up to number 70, or one numerical increment for larger main jets, until the acceleration time stops reducing. If there is no reduction in acceleration time through increasing the main jet sizes, start reducing the sizes of the main jets by one increment at a time and note what happens. If there is a reduction in time, the engine has responded to the slightly leaner setting, so continue to reduce main jet size one size at a time. The smallest main jet that causes the engine to accelerate the car in the quickest time possible is the size you are looking for. For short burst acceleration testing such as drag racing, stay with the minimum jetting. For circuit work, however, use slightly over-rich main jetting. Slightly rich mixtures do tend to assist in keeping the engine cooler, and this can be quite an advantage in endurance racing.

It may well be that the Holley recommended main jet sizes are absolutely correct for your application, but you can only really ascertain this by checking and testing with larger and then, if necessary, smaller main jets in the secondary barrels.

The main jets or metering plate mains drillings are the secondary barrels' overall mixture control adjustment system. On a vacuum secondary carburetor, or a mechanical secondary carburetor without a power valve, the main jets control the total amount of fuel that can go into the engine via the secondary barrels. On a vacuum secondary carburetor which has a power valve, or a mechanical secondary carburetor which has a power valve, the main jets and the power valve channel restriction holes supply the total amount of fuel that goes into the engine via the secondary barrels. The main jets are the main adjustment factor in all cases.

TUNING VACUUM SECONDARY CARBURETOR SECONDARY BARRELS

Follow the directions in the 'Using acceleration timing to optimise jetting' section of this chapter to ascertain whether a change or a setting is suitable. Making one change at a time and testing the engine is a well founded principle. Although time consuming, it does show the effects of each and every change you make. Testing after multiple changes will only lead to confusion.

Changing main jet sizes (metering plate carburetors)

You can change the metering plate for a richer or leaner main jet one by obtaining one or more plates after consulting the 'Holley Secondary Metering Plate' chart in the *Holley Performance Parts Catalog*. Alternatively, if a richer mixture is required, you can drill the main jet passages of the existing metering plate to match the larger metering plate jet sizes specified in the 'Holley Secondary Metering Plate' list. Initially increase main jet sizes by two increments between tests to see how the engine responds to a progressively richer mixture. Using this method, the main jet sizes will ultimately be made too large (at the point when acceleration time starts increasing), then another metering plate of the same type has to be found and its main jet passages drilled to the known optimum size.

Changing main jet sizes (metering block carburetors)

One of the main reasons for using a secondary metering block over a metering plate is that main jets are very easy to change in metering blocks. Note that there is a conversion chart in the *Holley Performance Parts Catalog* which gives direct equivalents of metering blocks to metering plates.

Power valves

With a power valve in the metering block there are two elements of mixture control that can be altered. The main jets alone are supplying

TUNING THE SECONDARY BARRELS OF 4150 & 4160 CARBS

the fuel until manifold vacuum drops below the rated amount of the particular power valve. At this point the power valve opens and fuel is able to flow into the main jet well and richen the mixture. This system meters the fuel supply very precisely to the engine loading and offers a different fuel supply curve to non power valve-equipped metering blocks. Having a power valve system in the secondary metering block can make the engine much more 'crisp' in its responses to the throttle. It takes extra work to get the power valve rating absolutely correct, but it can make a difference to the overall engine response and performance if the power valve fuel supply is tailored to suit the application exactly. Carburetors with this feature are more adjustable than others.

Wide open throttle, full power mixture is adjusted by increasing or decreasing the size of the main jets until optimum acceleration is achieved. The point in the rpm range where the power valve opens is decided by the vacuum (Hg) rating of the particular valve. A high number rated power valve (9.5-10.5Hg) will open the instant the vacuum drops below the rated figure ($^{3}/_{4}$ throttle), while a low number rated power valve (2.5-3.5Hg) will only open when the inlet manifold vacuum drops below the rated figure (near wide open throttle). Before the power valve opens, the secondaries will be operating on the mains only (not a full power mixture). This factor can be very significant for part throttle work, in a sweeping turn, for example, keeping the mixture leaner than it otherwise would be. Fit the power valve that causes the engine to go to a full power mixture when and only when the engine needs it. On part throttle engines rarely need a full power mixture, which is precisely why Holley fitted the secondary power valve to some models of carburetor in the first place. This sort of vacuum secondary carburetor is the top of the line model.

Selecting vacuum secondary diaphragm springs

The firmest diaphragm spring (black) is the slowest opening one and it pays to try this spring after the stock recommended one has been tested (unless it is the stock recommended one). With the slowest opening diaphragm spring acceleration test time known as well as that of the standard spring, the effect of fitting a weaker diaphragm spring can be measured. Acceleration timing is the way to pick the best possible spring. The secondaries need to be opened at the right time and at the right rate to allow quickest acceleration. Usually one spring will produce the best acceleration. Trying all of the diaphragm springs in acceleration testing is a good idea as it's quite surprising sometimes as to what spring works best.

Note that black plastic vacuum secondary diaphragm housings (part number 20-59) are available which have detachable tops on them for quick and easy spring changes. Two screws are undone and then the top comes off and the spring is simply unclipped and changed. This is a good system as it prevents possible damage to the diaphragm as the mechanism does not have to be taken to pieces each time the spring is changed.

The 'signal' to open the vacuum secondary barrels comes from the primary barrels. Once there is sufficient vacuum to lift the check ball, the vacuum is fed to the diaphragm assembly causing the secondaries to open. They'll open at a rate which is controlled by the spring tension. This means that unlike a mechanically operated four barrel carburetor, the secondaries will not open until the back two barrels are definitely able to be correctly utilised. There are most definitely instances where vacuum secondary carburetors can offer better engine response than mechanical secondary carburetors.

Caution! There's one aspect of vacuum secondary Holley carburetors of which you need to be aware. On a car with a manual gearshift, shutting off the throttle to make quick gearshifts can cause a problem. What happens is that while it may appear that the throttle shuts instantly, in fact there is a slight delay caused by the diaphragm. In this split second, a well modified and responsive engine will tend to rev higher. Depending on the engine, over-revving and bent valves are a possibility. Obviously at the point of shifting gear, the engine will be at or near maximum rpm and to have the revs rise at all at this point is not acceptable. This potential problem needs to be considered carefully and is one of the reasons why, for racing applications, mechanical secondary carburetors are almost always used because they do shut off instantly. Fitting a rev limiter and the strongest throttle return springs that can sensibly be used are options that need to be considered if a vacuum secondary carburetor is going to be used in a competition environment.

TUNING 'DOUBLE PUMPER' OR MECHANICAL CARBURETOR SECONDARY BARRELS

With the primary and secondary barrels in operation, the Holley recommended stock jetting in the secondary side of the carburetor, the float levels checked, the secondary accelerator pump actuating lever adjusted exactly the same way as the primary accelerator pump, the engine should be acceleration tested. Nine times out of ten it will accelerate well with the standard recommended secondary jetting, but it might not accelerate as well as it could. With the baseline figures obtained from testing with the primary barrels only in operation, good conclusions can be made as to whether the engine's overall performance is better, or worse, when the secondary barrels are brought into operation using the suggested testing regime.

There are several aspects of secondary barrel tuning to take into consideration. The main jets must be sized correctly to supply a rich enough mixture to generate maximum torque (the stock recommended main jetting is virtually guaranteed to be reasonably correct, meaning that the engine will run well enough to allow the accelerator pump settings to be optimised).

The accelerator pump discharge nozzle must be the right size. The accelerator pump cam must be suitable and in the right position, and the accelerator pump must be large enough. However, don't use a 50cc pump if a 30cc one is capable of delivering enough fuel. The rate of delivery of fuel and the amount of fuel delivered by the pump must both be right. The requirement here is to supply just enough fuel to prevent any hesitation during the acceleration phase but not to over supply and cause the engine to have to burn off the excess fuel before it can develop decent torque and brake horsepower.

Secondary idling system

ALL Holley four barrel carburetors (vacuum or mechanical secondary) have secondary idling circuitry in them, but it's usually of a fixed nature with no adjustment possible. The secondary barrel idle discharge holes are under the progression idle slots in the rear barrels (one small hole per barrel). They can, in some instances, be quite hard to see as they are so small, but they're there. The reason for having a secondary idling system is to keep the fuel moving through the secondary fuel bowl even if the secondary barrels are not being used, or little used, and also to ensure that the correct fuel level is maintained. If a car is driven over rough terrain, for example, and the secondary fuel bowl becomes over full, it will have its level reduced via the secondary idle circuitry reasonably quickly.

'Four corner' idle mixture adjustment

Some mechanical four barrel carburetors have adjustable secondary idle circuits. When adjusting these carburetors, set the front and rear metering block idle adjustment screws each to one full turn out from the lightly seated position. Start the engine and set the idle speed to suit the type of camshaft fitted. Generally, all four idle adjustment screws need to be adjusted out a similar number of turns. However, in some instances, the type of inlet manifold used may require slightly different settings across the carburetor to achieve optimum idle smoothness and progression smoothness (throttle response) from just off idle. All four idle screws must be off their seats. The adjustable secondary idle mixture can assist with secondary barrel opening and progression because the secondary barrels' air/fuel mixture can be adjusted which is not possible with a fixed idle secondary system. It's another tuning option.

Accelerator pump settings

In the first instance fit the stock (standard) recommended accelerator pump, the correct nozzle size for the particular carburetor and, with the accelerator pump cam in either the 1st or 2nd position, adjust the accelerator pump actuating lever system for zero lash (free play), the same way that a

Secondary accelerator pump cam and the two adjustment holes at 1 and 2 (B). Note the two fuel level sight plugs at A and A. The choke tower of this carburetor has been removed and the front of the carburetor is on the right.

TUNING THE SECONDARY BARRELS OF 4150 & 4160 CARBS

Primary accelerator pump and linkage (B), main throttle linkage with the cam position holes 1 and 2 (A and A), and the secondary barrel link arm (C).

Underside view. The primary barrels are the bottom two barrels. The throttle, on the bottom left of this carburetor (C), activates the primary accelerator pump (A) and the rear barrels through a short linkage arm (D). The secondary throttle spindle transmits the throttle action across the carburetor to the secondary accelerator pump (E), which activates the secondary accelerator pump (B).

primary accelerator pump is adjusted. The idea is to adjust up and down movement out of the lever mechanism, but not by so much that there is no easy sideways movement of this lever.

Just like the primary side of the carburetor, there are two positions per cam and a whole range of nozzles to go through. It is not always quite so easy to test the suitability of the secondary nozzle, accelerator pump cam, and hole settings, compared to the primary barrel settings.

The primary barrels of four barrel carburetors are almost always opened in advance of the secondary barrels (there are a few exceptions where all four barrels open simultaneously), so the engine is already accelerating when the secondaries are actuated, even if the time frame is only a fraction of a second.

With the stock secondary accelerator pump and cam fitted, the engine must be tested to see if the acceleration is as smooth as it was with just the primary barrels in operation. That's means no hesitation (not enough fuel), or flat spots in the acceleration rate while the engine clears excessive fuel.

Like the primary side, secondary side accelerator pump cams can be categorised to a certain degree, but the listing is quite loose because the cam actions are not *pro rata*. Some cams supply more fuel initially, while others pump progressively more fuel as the throttle is opened. The Holley secondary cam listing is based on full stroke delivery and, as a consequence, shows the individual cam's maximum fuel delivery capability, not the nature of its action. The accompanying list (A) of combinations for the cams is in order of 'full shot' size. This rating system is not all that relevant because it does not take into account the rate of fuel delivery per degree of throttle

A Cams in approximate order of maximum fuel delivery	B Cams in approximate order of least to maximum shot size.
1 - black	1 - pink
2 - pink	2 - black
3 - white	3 - red
4 - red	4 - white
5 - green	5 - green
6 - orange	6 - orange
7 - light blue	7 - brown
8 - brown	8 - light blue
9 - yellow	9 - yellow

spindle rotation from idle. The second chart (B) is almost always more useful for obtaining optimum acceleration phase engine performance. The secondary throttle opening is nearly as critical as that of the primary barrels if maximum acceleration efficiency is to be obtained.

The rating of the 9 cams on the basis of the rate/amount of the fuel delivery for the first 30 to 40 degrees of throttle rotation (the area where it matters most) is generally more useful. The list (B) is a fair but approximate representation of each cam's action in order of the least shot size to the maximum shot size. Changing from the number 1 hole position to the number 2 hole position will increase the shot size by a proportion of both charts.

Warning! Note that yellow and brown cams are for use with 50cc fuel pumps only. If these cams are used in conjunction with a 30cc fuel pump, the throttle will very likely jam in the fully open position the very first time it is fully opened. The consequences could be disastrous.

Optimising main jets

Change the main jets for leaner ones (in minus 1 size increment stages) and repeat the acceleration tests. If the acceleration time reduces, keep reducing the main jet size until the times stop reducing. Jet upwards in single jet size increments if going smaller than the starting point jets cause the acceleration time to increase.

Through this process find the smallest sized main jets that give the best acceleration times and the largest size main jets that give the same acceleration times. What this does is give you a small range of main jet sizing. For short duration speed work, such as drag racing, the leaner setting of main jetting can be used. For endurance work, such as circuit racing, it's usually advisable to fit richer jetting which will assist to a degree in keeping the engine as cool as possible: the disadvantage is that the engine will use more fuel to do the same amount of 'work'.

Caution! Backfiring or 'spitting back' through the carburetor is a definite sign of an over-lean mixture. Do not persist when this happens: increase the size of the mains until it stops (this assumes that the leanness is not caused by insufficient fuel pressure/volume).

If you have the opportunity, fitting a fuel pressure gauge visible to the driver for the duration of the initial testing and setting up is a good idea. Note that the stock Holley fuel filters in the fuel bowls will not last indefinitely before getting clogged. It's quite possible to have good fuel pressure (minimum of 4psi to a maximum of 7psi) but to have insufficient fuel flow because a filter is partially blocked. You can remove these standard filters and install a large easy-to-get-at aftermarket filter into the fuel line (after the fuel pump and just before the carburetor) instead: such filters should be changed annually.

The reason for timing acceleration over the $1/4$ mile, or over the rpm range of the engine in one gear over a set distance, is because although the engine may sound extremely good, all may not be as good as it could be. Data is needed for testing, and testing is very necessary. There is nothing expensive about the testing regime recommended here (the price of a stop watch), and it is very accurate (to a tenth of a second, in fact).

Warning! Don't ever test alone and always have a decent sized fire extinguisher handy in case of emergency. Anything can happen during testing and you need to be prepared.

The most significant feature of the tuning technique mentioned in this book is the division of the primary and the secondary barrels for tuning purposes. If an engine can't be made to run on the primary barrels correctly there is no hope of the engine being able to run successfully on all four barrels. This technique allows for the primary barrels to be fully and correctly tuned first, and then the secondary barrels are tuned.

Chapter 7
Inlet manifolds

There is a vast array of inlet manifolding available these days, with each type having been made to suit various applications, rpm and torque band ranges, and installation conditions. Many companies have made inlet manifolds of a certain configuration through developing an original idea, while others have expanded on what the automakers have come up with. The ultimate height of the carburetor and air cleaner is very often a prime consideration/limitation on many engine installations. The other consideration being the configuration and the quest for optimum distribution within a height requirement. Others are designed for optimum distribution without a height limitation (such as many 360 degree specialised racing inlet manifolds).

In most instances, aftermarket aluminum inlet manifolds, of either the 180 degree/dual plane design or 360 degree/single plane design, will work better on a high performance V8

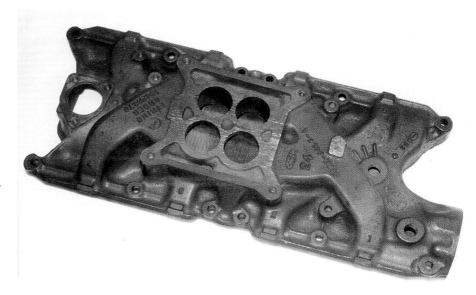

Typical 'negative rise' cast iron, 180 degree/dual plane, four barrel inlet manifold for a V8 engine.

engine than the standard cast iron 180 degree/dual plane inlet manifolds. The majority of aftermarket inlet manifolds of either type are 'high rising' types which will see the carburetor placed higher on the engine by around 1 to 3 inches on road going engine inlet manifolds and 7 to 9 inches on racing engine inlet manifolds. In each case, this is a desirable feature provided

SPEEDPRO SERIES

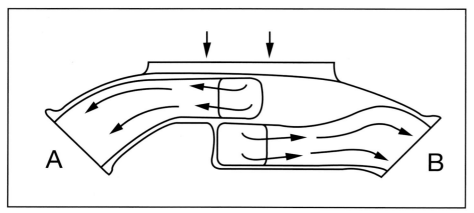

The bottom tier of inlet manifold runners at 'B' are not angled down into the cylinder head ports, unlike the inlet manifold runners at 'A'.

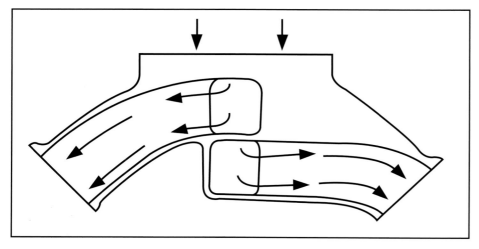

A 180 degree/dual plane 'high rise' type inlet manifold has both tiers of 'runners' angling down into the inlet ports of the cylinder heads, although the top tier is more acutely angled than the bottom.

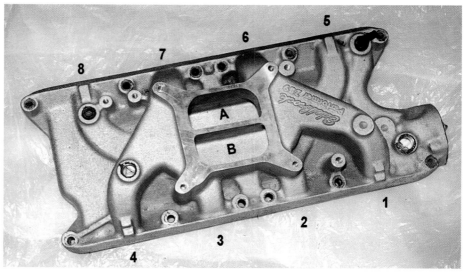

This aftermarket 180 degree/dual plane aluminium inlet manifold follows the original equipment manifold configuration quite closely, but it is 'high rise' whereas the original equipment cast iron one wasn't. Significantly, with these inlet manifolds, the carburetor barrels feeding into the manifold at A feed cylinders 2, 3, 5 and 8, while the carburetor barrels feeding into the manifold at B feed cylinders 1, 4, 6 and 7.

the inlet manifold, carburetor and air cleaner can be accommodated under the hood (bonnet). Within limits, the higher the manifold the better. Many modern specialist 360 degree/single plane racing inlet manifold runners are in line with the inlet ports making the carburetor sit very high.

'High rise' 180 degree/dual plane inlet manifolds allow much better routing, shaping and sizing of the manifold runners than most stock-type 180 degree manifolds, which are usually designed with engine compactness in mind. A 'high rise' manifold will usually improve engine acceleration characteristics, though brake horsepower may not be significantly affected.

While many inlet manifolds might look the same at a glance, in actual fact they are not. What fits often limits the choice but if height restrictions are not a consideration the choice of configuration can make a huge difference to the application. As a general rule, 180 degree/dual plane inlet manifolds are best for all 'low rpm' applications (off idle to 5500-6500 rpm) while the 360 degree/single plane inlet manifolds

INLET MANIFOLD

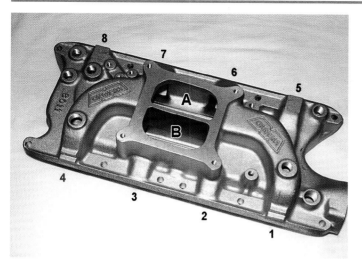

On this 180 degree/dual plane inlet manifold, the carburetor barrels that feed into the manifold at A feed cylinders 1, 4, 6 and 7, while the carburetor barrels that feed into the manifold at B feed cylinders 2, 3, 5 and 8. This is quite different to the preceding configuration. The difference being that the inlet manifold runners are not of equal length.

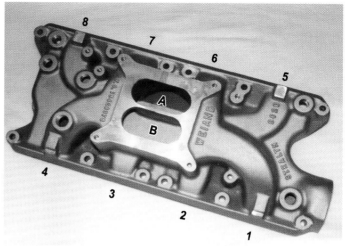

On these 180 degree/dual plane, 'high rise' inlet manifolds, the carburetor barrels that feed in at A feed cylinders 2, 3, 5 and 8, while the carburetor barrels that feed in at B feed cylinders 1, 4, 6 and 7. This is quite a different configuration to the two previous ones. The advantage over them being near equal length runners and a smooth flowing configuration. These manifolds are ideal for up to 6000-6500rpm use in road and race applications, and very often out-perform all other inlet manifolds.

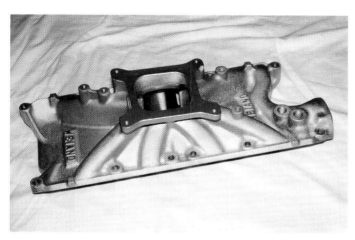

This 360 degree/single plane, 'high rise' inlet manifold has all four barrels feeding into a common area directly below the carburetor and all eight cylinders are fed off it. The individual runners all angle down into the inlet ports on the same level into the cylinder head and the shape, size and length of the runners are all designed to distribute the air fuel mixture as equally as possible. The individual runners are not however exactly equal length or the same shape.

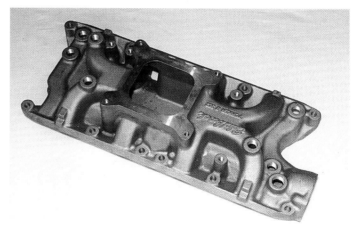

On this 360 degree/single plane inlet manifold all four barrels feed into a common area directly under the carburetor, and all eight cylinders feed off that. This inlet manifold is low for under-bonnet clearance. The carburetor is biased slightly towards the rear of the inlet manifold, in part to reduce the effect of the consecutive firing of the two adjacent rear cylinders.

www.velocebooks.com/www.veloce.co.uk
All books in print • New books • Special offers • Newsletter

SPEEDPRO SERIES

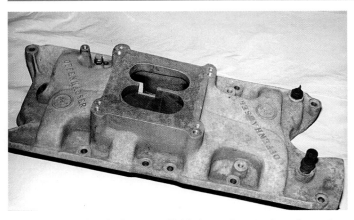

With this 360 degree inlet manifold, the carburetor barrels on the right feed the four cylinders on the right hand side of the engine, and the carburetor barrels on the left feed the cylinders on the left hand side of the engine. There is a slot in the divider which allows an amount of interaction between the two sides of the inlet manifold at low rpm. The carburetor is biased towards the rear of the inlet manifold to reduce the effect of the consecutive firing of the two adjacent rear cylinders.

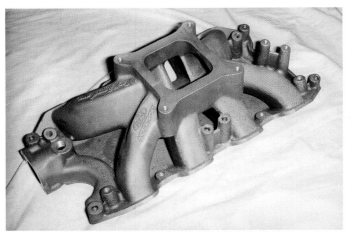

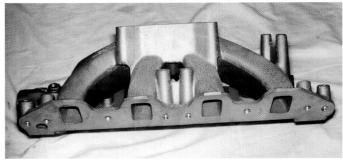

This is a 360 degree/single plane racing engine inlet manifold. The carburetor is mounted as high as practical and the individual runners are as sweeping as possible and near equal length.

generally favour high rpm applications but many of them adapt well to 'low rpm' applications (the small runner sized ones). Most inlet manifolds do as claimed by their respective manufacturers.

When it comes to 180 degree designs there are a few configurations available, although, unless you make a bit of a study of them you might miss the vital differences.

A good 180 degree/dual plane inlet manifold is ideal for engines that are revving to, but not exceeding 6500rpm, on a regular basis (probably the vast majority of engines being used today). A 'high rise' feature is most definitely beneficial if you're buying an aftermarket version of this type of inlet manifold. The cross sectional area of aftermarket inlet manifolds is frequently 10-20% larger than the standard ones that they replace. This design of inlet manifold is known for allowing good low and mid-range torque production.

360 degree/single plane inlet manifolds are the most popular (and best) for 6500rpm plus applications, and those where high revs are constantly required. These manifolds should have runners matching the cross sectional size of the head ports and the runners should be smooth flowing.

Manifold prices are reasonable, especially so for secondhand high performance manifolds. Inlet manifolds don't wear, although they do corrode around water inlets and outlets if the cooling system has not had a rust inhibitor/anti-freeze mixture added to the coolant.

The smaller the cross sectional area of the individual inlet manifold runners (provided they are at least the same size as the inlet ports), the higher the gas speed and the better the low rpm engine response. Engine acceleration capability determines what is good and what is bad when it comes to manifolds!

Since there are so many inlet manifolds available, and as they're not difficult to change, it's quite feasible to try a range of different models if you are really after the maximum performance from your engine.

www.velocebooks.com/www.veloce.co.uk
All books in print • New books • Special offers • Newsletter

Chapter 8
Aftermarket carburetor components

WEBER CARBURETOR AIR/FUEL METERING FOR HOLLEY CARBURETORS

The main jet, emulsion tube, air corrector and idle jet system of the DCOE and IDA Weber carburetor is legendary. The Holley two and four barrel carburetors are also 'fixed venturi' carburetors, just like the Webers, and the technology of the Weber can, as a consequence, be applied to them. Conversion kits have been available for years to do this very thing to the 2300 two barrel Holley carburetors and the primary barrels of 4150 and 4160 four barrel Holley carburetors. The Weber conversion kit comes with two individual 'blocks', a gasket, and four extra long securing screws, with the rest of the componentry required to complete the assembly being Holley parts.

Provided you are able to make all of the adjustments correctly, this system cannot be beaten on the basis of achieving a perfect air/fuel mixture for all engine conditions. There are, however, a huge number of possible combinations with this system, and that can be a problem in itself. A complete understanding of tuning Weber carburetors is required to be able to make adjustments that are worthwhile. Holley never went down this route on 2300, 4150 and 4160 carburetors, in

The Weber 'metering block' on the left comprises the power valve, idle adjustment screws and everything else that the Holley primary metering block houses except the idle jetting and the main jetting. The Weber idle and main jetting is housed in a separate Weber 'jetting block' (centre). A stock Holley 'centre pivot' fuel bowl is on the right, with the Weber-made, extra long securing screws fitted into it.

A Weber 'jetting block', with a Weber idle jet and idle jet holder below it on the left, and a Weber main jet, emulsion tube and air corrector on the right.

the vast majority of the calibration of the carburetors correct, but there is often some room for improvement. The Weber conversion can offer this improvement.

SECONDARY METERING PLATES

The stock Holley secondary 'metering plate', as used on millions of these carburetors, was the method used by the factory to reduce the cost of the carburetor without compromising efficiency. The metering plate does exactly as designed/required, but the rear main jetting is fixed. Holley makes a range of metering plates that cover every mixture and, if the existing range doesn't, abother metering plate could be added that did. All of the passageways and drilled holes can be individualised to suit any given application. The current range of Holley-made metering plates is, therefore, comprehensive, but it can mean that the end user might have to buy quite a few to be able to get the

Aftermarket secondary metering plate with removable Holley main jets.

rear venturi air/fuel mixture correct for his application.

To this end, machined aluminium secondary metering plates are available which allow the main jet portion of the fueling to be altered by the substitution of conventional Holley jets. The secondary idle feed holes will usually be suitable for most engines. At worst, they might have to be made smaller which will mean removing the original brass sleeves and pressing in new ones with smaller diameter holes in them. For the most part, you won't have to do this, and all that will have to be done is change the main jets to achieve the optimum secondary barrel air/fuel mixture ratio. These aftermarket, main jet substitution secondary metering plates can be excellent/convenient.

Note that it is possible to drill a stock Holley metering plate to take stock Holley main jets in the same manner. The only problem with doing this is that the stock Holley secondary metering plate is quite thin in this area, and care is needed when tapping the aluminium to make sure that the thread length is maximised. No matter how well the material is tapped, there is only ever going to be so much thread, and the main jet will only be being held in place by a few threads anyway. If the main jet is ever over-tightened, the thread in the aluminium will probably be stripped. The thread depth is ideal in the aftermarket metering plates and this will not happen in normal use.

an effort to keep the adjustments as few and as simple as possible for end users. Holley, for the most part, gets

Assembled Weber conversion on this 4160 carburetor. The Weber idle jets, main jet and emulsion tubes can be substituted without fully dismantling the carburetor.

TO ADVERTISE IN THIS SPACE CALL
0044 1305 260068
OR EMAIL INFO@VELOCE.CO.UK

APPENDIX - HOLLEY CARBURETOR NUMERICAL LISTINGS

Carburetor Part No.	Carb Model No.	CFM	Renew Kit	Trick Kit	Primary & Secondary Needle & Seat	Primary Main Jet	Secondary Main Jet or Plate	Primary Metering Block	Secondary Metering Block	Primary Power Valve	Primary Discharge Nozzle Size
0-1848-1	4160	465	37-119	37-933	6-506	122-57	N/S	N/S	34R9716-3	125-85	.025
0-1849	4160	550	37-119	37-933	6-506	122-62	N/S	N/S	N/S	125-85	.025
0-1850-2	4160	600	37-119	37-933	6-506	122-66	134-9	134-128	134-9	125-65	.025
0-1850-3	4160	600	37-119	37-933	6-506	122-66	134-9	134-128	134-9	125-65	.025
0-1850-4	4160	600	37-119	37-933	6-506	122-66	134-9	134-128	134-9	125-65	.031
0-1850-5	4160	600	37-119	37-933	6-506	122-66	134-9	134-128S	134-9	125-65	.031
0-2818-1	4150	600	37-1537	37-933	6-506	122-65	122-76	34R4094AS	N/S	125-65	.025
0-3124	4150	750	37-1539	37-933	6-504	122-70	122-76	N/S	N/S	125-85	(12) .025
0-3247	4150	780	37-1539	37-933	6-504	122-70	122-76	N/S	N/S	125-85	(12) .021
0-3259-1	4150	725	1085-2489	N/A	N/S	122-68	122-78	N/S	N/S	125-85	.025
0-3310-1	4150	780	37-1539	37-933	6-504	122-72	122-76	134-131	N/S	(12,13)	.025
0-3310-2	4160	750	37-754	37-933	6-504	122-72	134-21	134-131	134-21	125-65	.025
0-3310-3	4160	750	37-754	37-933	6-504	122-72	134-21	134-131	134-21	125-65	.025
0-3310-4	4160	750	37-754	37-933	6-504	122-72	134-21	134-131	134-21	125-65	.031
0-3310-5	4160	750	37-754	37-933	6-504	122-72	134-21	134-131	134-21	125-65	.031
0-3310-6	4160	750	37-754	37-933	6-504	122-72	134-21	134-131	134-21	125-65	.031
0-3367	4160	585	37-119	37-933	N/S	122-65	34R9716-22	N/S	N/R	125-65	.025
0-3370	4160	585	37-119	37-933	6-504	122-65	N/S	N/S	N/R	125-65	.025
0-3418-1	4150	855	37-1539	37-933	6-504	78C/82T	82C/80T	N/S	N/S	(15,21)	.025
0-3613	4150	770	37-1539	37-933	6-504	122-71	122-76	N/S	N/S	125-85	(12) .021
0-3659	2300	466	37-1537	37-933	6-504	N/R	N/S	N/R	N/S	N/R	N/R
0-3660	2300	350	37-1537	37-933	6-504	122-64	N/R	N/S	N/R	125-65	.021
0-3807	4150	595	37-1537	37-933	N/S	122-67	122-72	N/S	N/S	125-65	.025
0-3810	4160	585	37-1537	37-933	N/S	122-65	34R9716-22	N/S	N/R	125-65	.025
0-3811	4160	585	37-1537	37-933	N/S	122-65	N/S	N/S	N/R	125-65	.025
0-3910	4150	780	37-1539	37-933	6-504	122-71	122-76	N/S	N/S	125-65	(12) .021
0-4053	4150	780	37-1539	37-933	6-504	122-68	122-76	N/S	N/S	125-65	(12) .025
0-4055-1	2300	350	37-1537	37-933	6-504	122-63	N/R	N/S	N/R	125-65	.021
0-4056-1	2300	350	37-1537	37-933	6-504	122-61	N/R	N/S	N/R	125-65	.025
0-4118	4150	725	37-1539	37-933	6-504	122-68	122-78	N/S	N/S	125-85	.025
0-4144-1	2300	350	37-1537	37-933	6-504	122-62	N/R	N/S	N/R	125-65	.031
0-4224	4160	660	37-1537	37-933	6-508	122-76	34R9716-12	34R5913AS	34R9716-12	N/R	.025
0-4235	4160	770	37-485	37-933	6-504	(29)	N/S	N/S	N/R	125-65	.035
0-4236	4160	770	37-485	37-933	6-504	122-80	N/S	N/S	N/S	125-65	.035
0-4295	4150	585	37-485	37-933	6-504	122-69	122-71	N/S	N/S	125-65	.025
0-4296	4150	850	37-485	37-933	6-504	78C/82T	82C/80T	N/S	N/S	125-65	(15) .035
0-4346	4150	780	37-1539	37-933	6-504	122-68	122-76	N/S	N/S	125-85	(12) .025
0-4365-1	2300	500	37-1537	37-933	6-504	N/S	N/S	N/R	N/R	N/R	N/R
0-4412	2300	500	37-474	37-933	6-504	122-73	N/R	134-137	N/R	125-50	.028
0-4412-1	2300	500	37-474	37-933	6-504	122-73	N/R	134-137	N/R	125-50	.028
0-4412-2	2300	500	37-474	37-933	6-504	122-73	N/R	134-137	N/R	125-50	.028
0-4412-3	2300	500	37-474	37-933	6-504	122-73	N/R	134-137	N/R	125-50	.028
0-4452-1	4160	600	37-119	37-933	6-506	122-63	134-39	N/S	N/S	125-85	.031
0-4490	4150	780	37-1539	37-933	6-504	122-70	122-76	N/S	N/S	125-85	(12) .025
0-4514-1	4150	700	37-1537	37-933	6-504	122-66	122-79	N/S	N/S	125-65	.029
0-4548	4160	450	37-119	37-933	6-506	122-57	N/S	N/S	N/S	N/S	.031
0-4555	4150	780	37-1539	37-933	6-504	122-70	122-76	N/S	N/S	125-85	(12) .025
0-4575	4500	1050	37-1539	37-933	6-504	122-84	122-84	N/S	N/S	125-65	(15) .035
0-4609	4150	730	37-1537	37-933	6-504	122-66	122-79	N/S	N/S	125-65	.029

© **HOLLEY CARBURETORS**
VISIT <WWW.HOLLEY.COM> FOR CURRENT VERSIONS

Secondary Nozzle Size or Spring Color	Primary Bowl Gasket†	Primary Metering Block Gasket†	Secondary Bowl Gasket†	Secondary Metering Block Gasket†	Secondary Metering Plate Gasket†	Primary Fuel Bowl	Secondary Fuel Bowl	Throttle Body & Shaft Assembly	Venturi Diameter Primary	Venturi Diameter Secondary	Throttle Bore Diameter Primary	Throttle Bore Diameter Secondary
Green	108-83-2	108-89-2	108-90-2	108-90-2	108-27-2	34R2456A	134-105	N/S	1-3/32	1-3/32	1-1/2	1-1/2
Plain	108-83-2	108-89-2	108-90-2	108-90-2	108-27-2	34R2456A	134-105	N/S	1-3/16	1-1/4	1-1/2	1-1/2
Plain	108-83-2	108-89-2	108-90-2	108-90-2	108-27-2	134-101	134-105	112-20	1-1/4	1-5/16	1-9/16	1-9/16
Plain	108-83-2	108-89-2	108-90-2	108-90-2	N/R	134-101	134-105	112-20	1-1/4	1-5/16	1-9/16	1-9/16
Plain	108-83-2	108-89-2	108-90-2	108-90-2	N/R	134-101	134-105	112-20	1-1/4	1-5/16	1-9/16	1-9/16
Plain	108-83-2	108-89-2	108-90-2	108-90-2	N/R	134-101S	134-105S	112-20	1-1/4	1-5/16	1-9/16	1-9/16
Purple	108-83-2	108-89-2	108-83-2	108-89-2	N/R	134-101	34R5987AQ	N/S	1-1/4	1-5/16	1-9/16	1-9/16
Yellow	108-83-2	108-89-2	108-83-2	108-89-2	N/R	N/S	N/S	N/S	1-3/8	1-7/16	1-11/16	1-11/16
Yellow	108-83-2	108-89-2	108-83-2	108-89-2	N/R	N/S	134-102	N/S	1-3/8	1-7/16	1-11/16	1-11/16
Yellow	108-83-2	108-89-2	108-83-2	108-89-2	N/R	N/S	N/S	N/S	1-5/16	1-3/8	1-11/16	1-11/16
Plain	108-83-2	108-89-2	108-83-2	108-89-2	N/R	134-103	134-102	112-9	1-3/8	1-7/16	1-11/16	1-11/16
Plain	108-83-2	108-89-2	108-90-2	108-90-2	108-27-2	134-103	134-102	112-9	1-3/8	1-7/16	1-11/16	1-11/16
Plain	108-83-2	108-89-2	108-90-2	108-90-2	108-27-2	134-103	134-102	112-9	1-3/8	1-7/16	1-11/16	1-11/16
Plain	108-83-2	108-89-2	108-90-2	108-90-2	108-27-2	134-103S	134-102S	12R11311AP	1-3/8	1-7/16	1-11/16	1-11/16
Plain	108-83-2	108-89-2	108-90-2	108-90-2	108-27-2	134-103S	134-102S	12R11311AP	1-3/8	1-7/16	1-11/16	1-11/16
Purple	108-83-2	108-89-2	108-90-2	108-90-2	108-27-2	N/S	134-105	N/S	1-1/4	1-5/16	1-9/16	1-9/16
Purple	108-83-2	108-89-2	108-90-2	108-90-2	108-27-2	N/S	134-105	N/S	1-1/4	1-5/16	1-9/16	1-9/16
Yellow	108-83-2	108-89-2	108-83-2	108-89-2	N/R	134-103	N/S	N/S	1-9/16	1-9/16	1-3/4	1-3/4
Yellow	108-83-2	108-89-2	108-83-2	108-89-2	N/R	N/S	N/S	N/S	1-3/8	1-7/16	1-11/16	1-11/16
Brown	N/R	N/R	108-90-2	108-90-2	108-27-2	N/R	N/S	N/S	1-3/8	N/A	1-3/4	N/A
N/R	108-83-2	108-89-2	N/R	N/R	N/R	N/S	N/S	N/S	1-3/16	N/A	1-1/2	N/A
Purple	108-83-2	108-89-2	108-90-2	108-90-2	108-27-2	N/S	N/S	N/S	1-1/4	1-5/16	1-9/16	1-9/16
Purple	108-83-2	108-89-2	108-90-2	108-90-2	108-27-2	N/S	N/S	N/S	1-1/4	1-5/16	1-9/16	1-9/16
Yellow	108-83-2	108-89-2	108-83-2	108-89-2	N/R	134-103	N/S	N/S	1-3/8	1-7/16	1-11/16	1-11/16
Yellow	108-83-2	108-89-2	108-83-2	108-89-2	N/R	134-103	N/S	N/S	1-3/8	1-7/16	1-11/16	1-11/16
N/R	108-83-2	108-89-2	N/R	N/R	N/R	N/S	N/S	N/S	1-3/16	N/A	1-1/2	N/A
N/R	108-83-2	108-89-2	N/R	N/R	N/R	N/S	N/S	N/S	1-3/16	N/A	1-1/2	N/A
Yellow	108-83-2	108-89-2	108-83-2	108-89-2	N/R	134-103	134-102	N/S	1-5/16	1-3/8	1-11/16	1-11/16
N/R	108-83-2	108-89-2	N/R	N/R	N/R	N/S	N/R	N/S	1-3/16	N/A	1-1/2	N/A
.025	108-83-2	108-89-2	108-90-2	108-90-2	108-27-2	134-101	34R5987AQ	12R4280-3AM	1-1/4	1-5/16	1-11/16	1-11/16
Plain	108-83-2	108-89-2	108-90-2	108-90-2	108-27-2	134-101	34R5987AQ	N/S	1-3/8	1-7/16	1-11/16	1-11/16
Plain	108-83-2	108-89-2	108-90-2	108-90-2	108-27-2	134-101	34R5987AQ	N/S	1-3/8	1-7/16	1-11/16	1-11/16
.025	108-83-2	108-89-2	108-83-2	108-89-2	N/R	N/S	N/S	N/S	1-1/4	1-5/16	1-9/16	1-9/16
.035	108-83-2	108-89-2	108-83-2	108-89-2	N/R	134-103	N/S	N/S	1-9/16	1-9/16	1-3/4	1-3/4
Yellow	108-83-2	108-89-2	108-83-2	108-89-2	N/R	N/S	N/S	N/S	1-3/8	1-7/16	1-11/16	1-11/16
Yellow	N/R	N/R	108-90-2	108-90-2	108-13-2	N/S	N/R	N/S	1-9/16	N/A	1-3/4	N/A
N/R	108-83-2	108-89-2	N/R	N/R	N/R	134-103	N/R	112-2	1-3/8	N/R	1-11/16	N/R
N/R	108-83-2	108-89-2	N/R	N/R	N/R	134-103	N/R	112-2	1-3/8	N/R	1-11/16	N/R
N/R	108-83-2	108-89-2	N/R	N/R	N/R	134-103	N/R	112-2	1-3/8	N/R	1-11/16	N/R
N/R	108-83-2	108-89-2	N/R	N/R	N/R	134-103S	N/R	112-2	1-3/8	N/R	1-11/16	N/R
Purple	108-83-2	108-89-2	108-90-2	108-90-2	108-27-2	34R2456AQ	134-105	N/S	1-1/4	1-5/16	1-9/16	1-9/16
Yellow	108-83-2	108-89-2	108-83-2	108-89-2	N/R	134-103	N/S	N/S	1-3/8	1-7/16	1-11/16	1-11/16
Yellow	108-83-2	108-89-2	108-83-2	108-89-2	N/R	N/S	N/S	N/S	1-3/8	1-7/16	1-11/16	1-11/16
Brown	108-83-2	108-89-2	108-90-2	108-90-2	108-27-2	34R2456AQ	134-105	N/S	1-3/32	1-3/32	1-1/2	1-1/2
Yellow	108-83-2	108-89-2	108-83-2	108-89-2	N/R	134-103	N/S	N/S	1-3/8	1-7/16	1-11/16	1-11/16
.035	108-83-2	108-89-2	108-89-2	108-89-2	N/R	134-108	134-112	N/S	1-11/16	1-11/16	2	2
Yellow	108-83-2	108-89-2	108-83-2	108-89-2	N/R	N/S	N/S	N/S	1-3/8	1-7/16	1-11/16	1-11/16

APPENDIX - HOLLEY CARBURETOR NUMERICAL LISTINGS

Carburetor Part No.	Carb. Model No.	CFM	Renew Kit	Trick Kit	Primary & Secondary Needle & Seat	Primary Main Jet	Secondary Main Jet or Plate	Primary Metering Block	Secondary Metering Block	Primary Power Valve	Primary Discharge Nozzle Size
0-4628	4150	780	37-1537	37-933	6-504	122-70	122-83	N/S	N/S	125-85	.026
0-4647	4150	735	37-1537	37-933	6-504	122-64	122-82	N/S	N/S	125-85	.031
0-4653	4150	780	37-1537	37-933	6-504	122-71	122-82	N/S	N/S	125-65	.026
0-4670	2300	350	37-1537	37-933	6-504	122-62	N/R	N/S	N/R	125-65	.031
0-4672	2300	500	37-1537	37-933	6-504	N/R	N/S	N/R	N/R	N/R	N/R
0-4691-2	2110	300	N/A	N/A	N/S	122-63	N/R	N/R	N/R	N/R	.021
0-4742	4150	600	37-1539	37-933	6-504	122-63	122-72	N/S	N/S	N/S	.031
0-4776	4150	600	37-485	37-933	6-504	122-69	122-71	N/S	34R6502-3AM	125-65	.025
0-4776-1	4150	600	37-485	37-933	6-504	122-66	122-76	N/S	34R6502-3AM	125-65	.028
0-4776-2	4150	600	37-485	37-933	6-504	122-66	122-76	34R8519AS	34R6502-3AM	125-65	.028
0-4776-3	4150	600	37-485	37-933	6-504	122-66	122-73	34R8519AS	34R6502-3AM	125-65	.028
0-4776-4	4150	600	37-485	37-933	6-504	122-66	122-73	34R8519AS	34R6502-3AM	125-65	.028
0-4776-5	4150	600	37-485	37-933	6-504	122-66	122-73	34R8519AP	34R6502-3AMP	125-65	.028
0-4777	4150	650	37-485	37-933	6-504	122-71	122-76	N/S	34R6497AS	125-65	.025
0-4777-1	4150	650	37-485	37-933	6-504	122-67	122-76	N/S	34R6497AS	125-65	.028
0-4777-2	4150	650	37-485	37-933	6-504	122-67	122-76	134-150	34R6497AS	125-65	.028
0-4777-3	4150	650	37-485	37-933	6-504	122-67	122-73	134-150	34R6497AS	125-65	.028
0-4777-4	4150	650	37-485	37-933	6-504	122-67	122-73	134-150	34R6497AS	125-65	.028
0-4777-5	4150	650	37-485	37-933	6-504	122-67	122-73	34R8539-5AMP	34R6497-3AMP	125-65	.028
0-4778	4150	700	37-485	37-933	6-504	122-66	122-71	N/S	N/S	125-65	.025
0-4778-1	4150	700	37-485	37-933	6-504	122-66	122-76	N/S	N/S	125-65	.028
0-4778-2	4150	700	37-485	37-933	6-504	122-66	122-76	N/S	N/S	125-65	.028
0-4778-3	4150	700	37-485	37-933	6-504	122-69	122-78	N/S	N/S	125-65	.028
0-4778-4	4150	700	37-485	37-933	6-504	122-69	122-78	N/S	N/S	125-65	.028
0-4778-5	4150	700	37-485	37-933	6-504	122-69	122-78	34R11174AP	34R11176AP	125-65	.028
0-4779	4150	750	37-485	37-933	6-504	122-75	122-76	N/S	N/S	125-85	.025
0-4779-1	4150	750	37-485	37-933	6-504	122-70	122-80	N/S	N/S	125-85	.028
0-4779-2	4150	750	37-485	37-933	6-504	122-70	122-80	134-155	N/S	125-65	.028
0-4779-3	4150	750	37-485	37-933	6-504	122-70	122-73	34R11179AQ	N/S	125-65	.028
0-4779-4	4150	750	37-485	37-933	6-504	122-70	122-80	34R11179AQ	N/S	125-65	.028
0-4779-5	4150	750	37-485	37-933	6-504	122-70	122-80	34R11179AQ	N/S	125-65	.028
0-4779-6	4150	750	37-485	37-933	6-504	122-71	122-80	34R11179AQ	34R11041AQ	125-65	.028
0-4779-7	4150	750	37-485	37-933	6-504	122-71	122-80	34R11179APQ	34R11041APQ	125-65	.028
0-4780	4150	800	37-485	37-933	6-504	122-72	122-76	N/S	N/S	(12,21)	.031
0-4780-1	4150	800	37-485	37-933	6-504	122-70	122-76	N/S	N/S	(12,21)	.031
0-4780-2	4150	800	37-485	37-933	6-504	122-70	122-85	N/S	N/S	125-65	.031
0-4780-3	4150	800	37-485	37-933	6-504	122-71	122-85	34R11196AQ	N/S	125-65	.031
0-4780-4	4150	800	37-485	37-933	6-504	122-71	122-85	34R11196AQ	N/S	125-65	.031
0-4780-5	4150	800	37-485	37-933	6-504	122-71	122-85	34R11196APQ	34R11198APQ	125-65	.031
0-4781	4150	850	37-485	37-933	6-504	122-80	122-80	N/S	N/S	125-65	(15) .035
0-4781-1	4150	850	37-485	37-933	6-504	122-80	122-80	N/S	N/S	125-65	(15) .031
0-4781-2	4150	850	37-485	37-933	6-504	122-80	122-80	34R8558AS	N/S	125-65	(15) .031
0-4781-3	4150	850	37-485	37-933	6-504	122-80	122-78	N/S	N/S	125-65	(15) .031
0-4781-4	4150	850	37-485	37-933	6-504	122-80	122-78	N/S	N/S	125-65	(15) .031
0-4781-5	4150	850	37-485	37-933	6-504	122-80	122-78	34R11799AQ	N/S	125-65	(15) .031
0-4781-6	4150	850	37-485	37-933	6-504	122-80	122-78	34R11799APQ	34R9109-3AMPQ	125-65	(15) .031
0-4782	2300	355	37-1537	37-933	6-504	122-64	N/R	N/S	N/R	125-65	.031
0-4783	2300	500	37-1537	37-933	6-504	122-82	N/R	N/R	N/R	N/R	.028
0-4788	4150	830	37-485	37-933	6-504	122-80	122-80	N/S	N/S	125-65	(B) .031

HOLLEY CARBURETORS
VISIT <WWW.HOLLEY.COM> FOR CURRENT VERSIONS

Secondary Nozzle Size or Spring Color	Primary Bowl Gasket†	Primary Metering Block Gasket†	Secondary Bowl Gasket†	Secondary Metering Block Gasket†	Secondary Metering Plate Gasket†	Primary Fuel Bowl	Secondary Fuel Bowl	Throttle Body & Shaft Assembly	Venturi Diameter Primary	Venturi Diameter Secondary	Throttle Bore Diameter Primary	Throttle Bore Diameter Secondary
Yellow	108-83-2	108-89-2	108-83-2	108-89-2	N/R	N/S	N/S	N/S	1-3/8	1-7/16	1-11/16	1-11/16
Purple	108-83-2	108-89-2	108-83-2	108-89-2	N/R	N/S	N/S	N/S	1-3/8	1-7/16	1-11/16	1-11/16
Purple	108-83-2	108-89-2	108-83-2	108-89-2	N/R	N/S	N/S	N/S	1-3/8	1-7/16	1-11/16	1-11/16
N/R	108-83-2	108-89-2	N/R	N/R	N/R	N/S	N/R	N/S	1-3/16	N/A	1-1/2	N/A
Yellow	N/R	N/R	108-90-2	108-90-2	108-13-2	N/R	N/S	N/S	1-9/16	N/A	1-3/4	N/A
N/R	N/R	N/R	N/R	N/R	N/R	N/R	N/R	N/R	1-5/32	N/R	1-7/16	N/R
Purple	108-83-2	108-89-2	108-83-2	108-89-2	N/R	N/R	N/S	N/S	1-1/4	1-5/16	1-9/16	1-9/16
.032	108-83-2	108-89-2	108-83-2	108-89-2	N/R	134-103	134-104	N/S	1-1/4	1-5/16	1-9/16	1-9/16
.032	108-83-2	108-89-2	108-83-2	108-89-2	N/R	134-103	134-104	N/S	1-1/4	1-5/16	1-9/16	1-9/16
.032	108-83-2	108-89-2	108-83-2	108-89-2	N/R	134-103	134-104	12R11086A	1-1/4	1-5/16	1-9/16	1-9/16
.032	108-83-2	108-89-2	108-83-2	108-89-2	N/R	134-103	134-104	12R11086A	1-1/4	1-5/16	1-9/16	1-9/16
.032	108-83-2	108-89-2	108-83-2	108-89-2	N/R	134-103	134-104	12R11086A	1-1/4	1-5/16	1-9/16	1-9/16
.032	108-83-2	108-89-2	108-83-2	108-89-2	N/R	134-103S	134-104S	12R11086AP	1-1/4	1-5/16	1-9/16	1-9/16
.025	108-83-2	108-89-2	108-83-2	108-89-2	N/R	134-103	134-104	N/S	1-1/4	1-5/16	1-11/16	1-11/16
.028	108-83-2	108-89-2	108-83-2	108-89-2	N/R	134-103	134-104	N/S	1-1/4	1-5/16	1-11/16	1-11/16
.028	108-83-2	108-89-2	108-83-2	108-89-2	N/R	134-103	134-104	112-17	1-1/4	1-5/16	1-11/16	1-11/16
.028	108-83-2	108-89-2	108-83-2	108-89-2	N/R	134-103	134-104	112-17	1-1/4	1-5/16	1-11/16	1-11/16
.028	108-83-2	108-89-2	108-83-2	108-89-2	N/R	134-103	134-104	112-17	1-1/4	1-5/16	1-11/16	1-11/16
.028	108-83-2	108-89-2	108-83-2	108-89-2	N/R	134-103S	134-104S	12R11805AP	1-1/4	1-5/16	1-11/16	1-11/16
.032	108-83-2	108-89-2	108-83-2	108-89-2	N/R	134-103	134-104	N/S	1-5/16	1-3/8	1-11/16	1-11/16
.031	108-83-2	108-89-2	108-83-2	108-89-2	N/R	134-103	134-104	N/S	1-5/16	1-3/8	1-11/16	1-11/16
.031	108-83-2	108-89-2	108-83-2	108-89-2	N/R	134-103	134-104	12R11092A	1-5/16	1-3/8	1-11/16	1-11/16
.031	108-83-2	108-89-2	108-83-2	108-89-2	N/R	134-103	134-104	12R11092A	1-5/16	1-3/8	1-11/16	1-11/16
.031	108-83-2	108-89-2	108-83-2	108-89-2	N/R	134-103	134-104	12R11092A	1-5/16	1-3/8	1-11/16	1-11/16
.031	108-83-2	108-89-2	108-83-2	108-89-2	N/R	134-103S	134-104S	12R11092AP	1-5/16	1-3/8	1-11/16	1-11/16
.032	108-83-2	108-89-2	108-83-2	108-89-2	N/R	134-103	134-104	N/S	1-3/8	1-3/8	1-11/16	1-11/16
.031	108-83-2	108-89-2	108-83-2	108-89-2	N/R	134-103	134-104	N/S	1-3/8	1-3/8	1-11/16	1-11/16
.031	108-83-2	108-89-2	108-83-2	108-89-2	N/R	134-103	134-104	N/S	1-3/8	1-3/8	1-11/16	1-11/16
.031	108-83-2	108-89-2	108-83-2	108-89-2	N/R	134-103	134-104	N/S	1-3/8	1-3/8	1-11/16	1-11/16
.031	108-83-2	108-89-2	108-83-2	108-89-2	N/R	134-103	134-104	N/S	1-3/8	1-3/8	1-11/16	1-11/16
.031	108-83-2	108-89-2	108-83-2	108-89-2	N/R	134-103	134-104	12R11147AP	1-3/8	1-3/8	1-11/16	1-11/16
.031	108-83-2	108-89-2	108-83-2	108-89-2	N/R	134-103S	134-104S	12R11147AP	1-3/8	1-3/8	1-11/16	1-11/16
.031	108-83-2	108-89-2	108-83-2	108-89-2	N/R	134-103	134-104	N/S	1-3/8	1-7/16	1-11/16	1-11/16
.031	108-83-2	108-89-2	108-83-2	108-89-2	N/R	134-103	134-104	N/S	1-3/8	1-7/16	1-11/16	1-11/16
.031	108-83-2	108-89-2	108-83-2	108-89-2	N/R	134-103	134-104	12R11090AP	1-3/8	1-7/16	1-11/16	1-11/16
.031	108-83-2	108-89-2	108-83-2	108-89-2	N/R	134-103	134-104	12R11090A	1-3/8	1-7/16	1-11/16	1-11/16
.031	108-83-2	108-89-2	108-83-2	108-89-2	N/R	134-103	134-104	12R11090A	1-3/8	1-7/16	1-11/16	1-11/16
.031	108-83-2	108-89-2	108-83-2	108-89-2	N/R	134-103S	134-104S	12R11090AP	1-3/8	1-7/16	1-11/16	1-11/16
.025	108-83-2	108-89-2	108-83-2	108-89-2	N/R	134-103	134-104	N/S	1-9/16	1-9/16	1-3/4	1-3/4
.031	108-83-2	108-89-2	108-83-2	108-89-2	N/R	134-103	134-104	N/S	1-9/16	1-9/16	1-3/4	1-3/4
.031	108-83-2	108-89-2	108-83-2	108-89-2	N/R	134-103	134-104	N/S	1-9/16	1-9/16	1-3/4	1-3/4
.031	108-83-2	108-89-2	108-83-2	108-89-2	N/R	134-103	134-104	N/S	1-9/16	1-9/16	1-3/4	1-3/4
.031	108-83-2	108-89-2	108-83-2	108-89-2	N/R	134-103	134-104	12R11153AP	1-9/16	1-9/16	1-3/4	1-3/4
.031	108-83-2	108-89-2	108-83-2	108-89-2	N/R	134-103S	134-104S	12R11153AP	1-9/16	1-9/16	1-3/4	1-3/4
N/R	108-83-2	108-89-2	N/R	N/R	N/R	N/S	N/R	N/S	1-3/16	N/R	1-1/2	N/R
N/R	108-83-2	108-89-2	N/R	N/R	N/R	N/S	N/R	N/S	1-9/16	N/R	1-3/4	N/R
.031	108-83-2	108-89-2	108-83-2	108-89-2	N/R	134-103	134-104	N/S	1-9/16	1-9/16	1-11/16	1-11/16

APPENDIX - HOLLEY CARBURETOR NUMERICAL LISTINGS

Carburetor Part No.	Carb. Model No.	CFM	Renew Kit	Trick Kit	Primary & Secondary Needle & Seat	Primary Main Jet	Secondary Main Jet or Plate	Primary Metering Block	Secondary Metering Block	Primary Power Valve	Primary Discharge Nozzle Size
0-4788-1	4150	830	37-485	37-933	6-504	122-80	122-80	N/S	N/S	125-65	(B) .031
0-4790	2300	500	37-1537	37-933	6-504	N/R	N/S	N/R	N/R	N/R	N/R
0-4791	2300	350	37-1537	37-933	6-504	122-62	N/R	N/S	N/R	125-65	.031
0-4792	2300	350	37-1537	37-933	6-504	122-61	N/R	N/S	N/R	125-65	.031
0-4800-1	4150	780	37-1539	37-933	6-504	122-70	122-76	N/S	N/S	125-85	(12) .025
0-4801-1	4150	780	37-1539	37-933	6-504	122-70	122-76	N/S	N/S	128-85	(12) .025
0-4802-1	4150	780	37-1539	37-933	6-504	122-70	122-76	N/S	N/S	125-85	(12) .025
0-4803-1	4150	780	37-1539	37-933	6-504	122-70	122-76	N/S	N/S	125-85	(12) .025
0-6109	4150	750	37-485	37-933	6-504	122-75	122-76	N/S	N/S	125-85	.025
0-6129	4150	780	37-1537	37-933	6-504	122-70	122-82	N/S	N/S	125-65	.026
0-6210-1	4165	650	37-605	37-933	(16,17)	122-602	122-632	N/S	N/S	(14,15)	.025
0-6210-2	4165	650	37-605	37-933	(16,17)	122-602	122-83	N/S	N/S	125-85	.025
0-6210-3	4165	650	37-605	37-933	(16,17)	122-602	122-83	N/S	N/S	125-85	.025
0-6211	4165	800	37-605	37-933	(16,17)	122-62	122-85	N/S	N/S	(14,15)	.025
0-6211-1	4165	800	37-605	37-933	(16,17)	122-602	122-85	N/S	N/S	(14,15)	.025
0-6212	4165	800	37-1537	37-933	6-504	122-63	122-86	N/S	N/S	(14,15)	.025
0-6213	4165	800	37-1537	37-933	6-504	122-62	122-85	N/S	N/S	(14,15)	.025
0-6214	4500	1150	N/A	N/A	6-504	122-95	122-95	N/S	N/S	N/R	.026
0-6238-1	4150	780	37-1539	37-933	6-504	122-68	122-73	N/S	N/S	125-65	(12) .025
0-6239-1	4150	780	37-1539	37-933	6-504	122-68	122-73	N/S	N/S	125-65	(12) .025
0-6244-1	2110	200	N/A	N/A	6-509	122-47	N/R	N/R	N/R	N/R	.021
0-6262	4165	800	37-605	37-933	(16,17)	122-62	122-85	N/S	N/S	(14,15)	.025
0-6270-1	4160	600	37-1536	37-933	N/S	122-64	N/S	N/S	N/S	125-85	.032
0-6291	4160	600	37-119	37-933	6-506	122-62	134-39	N/S	N/S	125-85	.031
0-6299-1	4160	390	37-1539	37-933	6-506	122-50	34R9716-34	N/S	N/S	N/A	.025
0-6425	2300	650	N/A	N/A	6-504	122-82	N/R	N/S	N/S	125-65	.031
0-6464	4500	1050	37-1539	37-933	6-504	122-88	122-88	N/S	N/R	N/R	.035
0-6468-1	4165	650	37-605	37-933	(16,17)	122-60	122-83	N/S	N/S	125-85	.025
0-6468-2	4165	650	37-605	37-933	(16,17)	122-602	122-83	N/S	N/A	125-85	.025
0-6497	4165	650	37-605	37-933	(16,17)	122-582	122-602	N/S	N/S	(14,15)	.025
0-6498	4165	650	37-605	37-933	(16,17)	122-592	122-602	N/S	N/S	(14,15)	.025
0-6499	4165	650	37-1537	N/A	6-504	122-60	122-63	N/S	N/S	(14,15)	.025
0-6512	4165	650	37-605	37-933	(16,17)	122-60	122-60	N/S	N/S	(14,15)	.025
0-6520	4160	600	37-119	37-933	6-506	122-62	134-39	N/S	N/S	125-85	.031
0-6528	4165	650	37-605	37-933	(16,17)	122-61	122-60	N/S	N/S	(14,15)	.025
0-6619-1	4160	600	37-720	37-933	6-506	122-642	134-39	N/S	134-39	125-65	.031
0-6647	4150	600	3-655	N/A	6-504	122-68	122-70	N/S	N/S	125-85	(12) .025
0-6708	4150	650	37-1539	37-933	6-504	122-552	122-752	N/S	134-39	(21,22)	.025
0-6708-1	4150	650	37-1539	37-933	6-504	122-542	122-85	N/S	N/S	125-65	.025
0-6709	4150	750	37-1539	37-933	6-504	122-652	122-76	N/S	N/S	(21,22)	.025
0-6710	4165	800	37-1537	37-933	6-504	122-63	122-86	N/S	N/S	(21,22)	.025
0-6711	4165	650	37-605	37-933	(16,17)	122-602	122-632	N/S	N/S	(21,22)	.025
0-6772	4165	650	37-605	37-933	(16,17)	122-592	122-602	N/S	N/S	(14,15)	.025
0-6773	4165	650	37-605	37-933	(16,17)	122-592	122-602	N/S	N/S	(14,15)	.025
0-6774	4165	650	37-605	37-933	(16,17)	122-572	122-602	N/S	N/S	(14,15)	.025
0-6853	4165	650	37-605	37-933	(16,17)	122-60	122-62	N/S	N/S	(14,15)	.025
0-6895	4150	390	37-1536	37-933	6-504	122-50	122-62	N/S	N/S	125-85	.025
0-6909	4160	600	37-119	37-933	6-506	122-622	134-39	N/S	134-39	125-65	.031
0-6910	4165	800	37-1537	37-933	6-504	122-612	122-86	N/S	N/S	(14,15)	.025

© **HOLLEY CARBURETORS**
VISIT <WWW.HOLLEY.COM> FOR CURRENT VERSIONS

Secondary Nozzle Size or Spring Color	Primary Bowl Gasket†	Primary Metering Block Gasket†	Secondary Bowl Gasket†	Secondary Metering Block Gasket†	Secondary Metering Plate Gasket†	Primary Fuel Bowl	Throttle Secondary Fuel Bowl	Body & Shaft Assembly	Venturi Diameter Primary	Venturi Diameter Secondary	Throttle Bore Diameter Primary	Throttle Bore Diameter Secondary
.031	108-83-2	108-89-2	108-83-2	108-89-2	N/R	134-103	134-104	N/S	1-9/16	1-9/16	1-11/16	1-11/16
Yellow	N/R	N/R	108-90-2	108-90-2	108-13-2	N/R	N/S	N/S	1-9/16	N/R	1-3/4	N/R
N/R	108-83-2	108-89-2	N/R	N/R	N/R	N/S	N/R	N/S	1-3/16	N/R	1-1/2	N/R
N/R	108-83-2	108-89-2	N/R	N/R	N/R	N/S	N/R	N/S	1-3/16	N/R	1-1/2	N/R
Yellow	108-83-2	108-89-2	108-83-2	108-89-2	N/R	134-103	N/S	N/S	1-3/8	1-7/16	1-11/16	1-11/16
Yellow	108-83-2	108-89-2	108-83-2	108-89-2	N/R	134-103	N/S	N/S	1-3/8	1-7/16	1-11/16	1-11/16
Yellow	108-83-2	108-89-2	108-83-2	108-89-2	N/R	134-103	N/S	N/S	1-3/8	1-7/16	1-11/16	1-11/16
.032	108-83-2	108-89-2	108-83-2	108-89-2	N/R	N/S	N/S	N/S	1-3/8	1-3/8	1-11/16	1-11/16
Purple	108-83-2	108-89-2	108-83-2	108-89-2	N/R	N/S	N/S	N/S	1-3/8	1-7/16	1-11/16	1-11/16
.037	108-92-2	108-91-2	108-92-2	108-91-2	N/R	134-110	34R7201A	N/S	1-5/32	1-3/8	1-3/8	2
.037	108-92-2	108-91-2	108-92-2	108-91-2	N/R	134-110	34R7201A	N/S	1-5/32	1-3/8	1-3/8	2
.037	108-92-2	108-91-2	108-92-2	108-91-2	N/R	134-110	34R7201A	12R7222-4AM	1-5/32	1-3/8	1-3/8	2
.037	108-92-2	108-91-2	108-92-2	108-91-2	N/R	134-110	34R7201A	N/S	1-5/32	1-23/32	1-3/8	2
.037	108-92-2	108-91-2	108-92-2	108-91-2	N/R	134-110	34R7201A	N/S	1-5/32	1-23/32	1-3/8	2
.037	108-92-2	108-91-2	108-92-2	108-91-2	N/R	N/S	N/S	N/S	1-5/32	1-23/32	1-3/8	2
.026	108-83-2	108-36-2	108-83-2	108-36-2	N/R	134-108	134-112	N/S	1-13/16	1-13/16	2	2
Yellow	108-83-2	108-89-2	108-83-2	108-89-2	N/R	N/S	N/S	N/S	1-3/8	1-7/16	1-11/16	1-11/16
Yellow	108-83-2	108-89-2	108-83-2	108-89-2	N/R	N/S	N/S	N/S	1-3/8	1-7/16	1-11/16	1-11/16
N/R	N/R	N/R	N/R	N/R	N/R	N/R	N/R	N/R	1-5/16	N/R	1-7/16	N/R
.037	108-92-2	108-91-2	108-92-2	108-91-2	N/R	134-110	34R7201A	N/S	1-13/16	1-23/32	1-3/8	2
Orange	108-83-2	108-34-2	108-90-2	108-90-2	108-27-2	N/S	N/S	N/S	1-1/4	1-5/16	1-9/16	1-9/16
Purple	108-83-2	108-91-2	108-90-2	108-90-2	108-27-2	N/S	N/S	N/S	1-1/4	1-5/16	1-5/16	1-9/16
Plain	108-83-2	108-89-2	108-90-2	108-90-2	108-28-2	N/S	N/S	N/S	1-1/16	1-1/16	1-7/16	1-7/16
N/R	108-92-2	108-35-2	N/R	N/R	N/R	N/S	N/S	N/S	1-7/16	N/R	1-3/4	N/R
.035	108-83-2	108-36-2	108-83-2	108-36-2	N/R	134-108	134-112	N/S	1-11/16	1-11/16	2	2
.037	108-92-2	108-91-2	108-92-2	108-91-2	N/R	134-110	34R7201A	N/S	1-5/32	1-3/8	1-3/8	2
.037	108-92-2	108-91-2	108-92-2	108-91-2	N/R	134-110	34R7201A	N/S	1-5/32	1-3/8	1-3/8	2
.037	108-92-2	108-91-2	108-92-2	108-91-2	N/R	134-110	34R7201A	N/S	1-5/32	1-3/8	1-3/8	2
.037	108-92-2	108-91-2	108-92-2	108-91-2	N/R	N/S	N/S	N/S	1-5/32	1-3/8	1-3/8	2
.037	108-92-2	108-91-2	108-92-2	108-91-2	N/R	134-110	34R7201A	N/S	1-5/32	1-3/8	1-3/8	2
Purple	108-83-2	108-91-2	108-90-2	108-90-2	108-27-2	N/S	N/S	N/S	1-1/4	1-5/16	1-9/16	1-9/16
.037	108-92-2	108-91-2	108-92-2	108-91-2	N/R	134-110	34R7201A	N/S	1-5/32	1-3/8	1-3/8	2
Black	108-83-2	108-91-2	108-90-2	108-90-2	108-27-2	34R8242AQ	134-105	N/S	1-1/4	1-5/16	1-9/16	1-9/16
Yellow	108-83-2	108-89-2	108-83-2	108-89-2	N/R	N/S	N/S	N/S	1-1/4	1-5/16	1-9/16	1-9/16
.037	108-92-2	108-91-2	108-92-2	108-91-2	N/R	N/S	N/S	N/S	1-3/32	1-9/16	1-1/2	1-3/4
.037	108-92-2	108-91-2	108-92-2	108-91-2	N/R	N/S	N/S	N/S	1-3/32	1-9/16	1-1/2	1-3/4
.037	108-92-2	108-91-2	108-92-2	108-91-2	N/R	N/S	N/S	N/S	1-1/4	1-9/16	1-1/2	1-3/4
.037	108-92-2	108-91-2	108-92-2	108-91-2	N/R	N/S	N/S	N/S	1-5/32	1-23/32	1-3/8	2
.028	108-92-2	108-91-2	108-92-2	108-91-2	N/R	N/S	N/S	34R7201A	N/S	1-5/32	1-3/8	1-3/8	2
.040	108-92-2	108-91-2	108-92-2	108-91-2	N/R	134-110	34R7201A	N/S	1-5/32	1-3/8	1-3/8	2
.040	108-92-2	108-91-2	108-92-2	108-91-2	N/R	134-110	34R7201A	N/S	1-5/32	1-3/8	1-3/8	2
.037	108-92-2	108-91-2	108-92-2	108-91-2	N/R	134-110	34R7201A	N/S	1-5/32	1-3/8	1-3/8	2
.037	108-92-2	108-91-2	108-92-2	108-91-2	N/R	134-110	34R7201A	N/S	1-5/32	1-3/8	1-3/8	2
.025	108-83-2	108-89-2	108-83-2	108-89-2	N/R	134-103	134-104	N/S	1-1/16	1-1/16	1-7/16	1-7/16
Black	108-83-2	108-91-2	108-90-2	108-90-2	108-27-2	N/S	134-105	N/S	1-1/4	1-5/16	1-9/16	1-9/16
.037	108-92-2	108-91-2	108-92-2	108-91-2	N/R	N/S	N/S	N/S	1-5/32	1-23/32	1-3/8	2

APPENDIX - HOLLEY CARBURETOR NUMERICAL LISTINGS

Carburetor Part No.	Carb. Model No.	CFM	Renew Kit	Trick Kit	Primary & Secondary Needle & Seat	Primary Main Jet	Secondary Main Jet or Plate	Primary Metering Block	Secondary Metering Block	Primary Power Valve	Primary Discharge Nozzle Size
0-6919	4160	600	37-1536	37-933	6-506	122-622	134-39	N/S	134-39	125-206	.031
0-6946-1	4160	600	3-1012	N/A	6-504	122-612	N/S	N/S	N/S	125-211	.025
0-6947	4160	600	3-1012	N/A	6-504	122-612	N/S	N/S	N/S	125-206	.025
0-6979	4160	600	N/A	37-933	6-506	122-642	134-39	N/S	134-39	125-85	.031
0-6979-1	4160	600	N/A	37-933	6-506	122-642	134-39	N/S	134-39	125-208	.031
0-6989	4160	600	37-1536	37-933	6-506	122-622	134-39	N/S	134-39	125-206	.031
0-7001	4165	650	N/A	37-933	(16,17)	122-582	122-602	N/S	N/S	(15,24)	.025
0-7002-1	4175	650	37-1537	37-933	(16,17)	122-582	134-21	N/S	134-21	125-85	.025
0-7004-1	4175	650	37-1537	37-933	(16,17)	122-562	34R9716-45	N/S	34R9716-45	125-212	.025
0-7004-2	4175	650	37-1537	37-933	(16,17)	122-542	N/S	N/S	N/S	125-211	.025
0-7005-1	4175	650	37-1537	37-933	(16,17)	122-562	34R9716-45	N/S	34R9716-45	125-212	.025
0-7005-2	4175	650	37-1537	37-933	(16,17)	122-542	N/S	N/S	N/S	125-212	.025
0-7006-1	4175	650	37-1537	37-933	(16,17)	122-562	34R9716-45	N/S	34R9716-45	125-212	.025
0-7006-2	4175	650	37-1537	37-933	(16,17)	122-542	N/S	N/S	N/S	125-211	.025
0-7009-1	4160	600	37-1536	37-933	6-506	122-622	134-39	N/S	134-39	125-206	.031
0-7010	4160	780	37-1537	37-933	6-506	122-662	N/S	N/S	N/S	125-65	.025
0-7053-1	4160	600	37-119	37-933	6-506	122-632	134-39	N/S	N/S	125-85	.031
0-7054	4165	650	37-605	37-933	(16,17)	122-592	122-602	N/S	N/S	(14,15)	.025
0-7154	4160	600	37-119	37-933	6-506	122-62	N/S	N/S	N/S	125-85	.031
0-7320	4500	1150	37-1539	37-933	6-504	122-95	122-95	N/S	N/S	N/A	.031
0-7320-1	4500	1150	37-1539	37-933	6-518-2	122-95	122-95	N/S	N/S	N/R	.031
0-7343	5200	230	N/A	N/A	N/S	N/S	N/S	N/S	N/S	N/S	.020
0-7344	5210	255	N/A	N/A	N/S	N/S	N/S	N/S	N/S	N/S	.021
0-7351	4175	650	37-1537	37-933	(16,17)	122-592	N/S	N/S	134-21	125-206	.037
0-7397	4175	650	37-1537	37-933	(16,17)	122-582	134-21	N/S	134-21	125-206	.037
0-7410	4150	340	37-1536	37-933	6-504	122-50	122-62	N/S	N/S	125-85	.025
0-7411	4150	370	37-1536	37-933	6-504	122-50	122-62	N/S	N/S	125-85	.025
0-7413	4160	600	37-119	37-933	6-506	122-632	134-39	N/S	N/S	125-85	.031
0-7448	2300	350	37-1536	37-933	6-504	122-61	N/A	134-203	N/R	125-85	.031
0-7454	4360	450	37-1540	N/A	N/S	124-215	124-550	N/R	N/R	N/S	.028
0-7455	4360	450	37-1540	N/A	N/S	124-215	124-537	N/R	N/R	N/S	.028
0-7456	4360	450	37-1540	N/A	N/S	124-215	124-550	N/R	N/R	N/S	.028
0-7555	4360	450	37-1540	N/A	N/S	124-215	124-550	N/R	N/R	N/S	.028
0-7556	4360	450	37-1540	N/A	N/S	124-215	124-550	N/R	N/R	N/S	.028
0-7850	4160	600	N/A	N/A	6-506	122-622	134-39	N/S	N/S	125-85	.031
0-7855	4175	650	37-1537	37-933	(16,17)	122-562	34R9716-45	N/S	34R9716-45	125-212	.028
0-7955	4360	450	37-1540	N/A	N/S	124-219	124-550	N/R	N/R	N/S	.028
0-7956	4360	450	37-1540	N/A	N/S	124-239	124-550	N/R	N/R	N/S	.028
0-7957	4360	450	37-1540	N/A	N/S	124-219	124-550	N/R	N/R	N/S	.028
0-7958	4360	450	37-1540	N/A	N/S	124-219	124-550	N/R	N/R	N/S	.028
0-7985	4160	600	37-1536	37-933	6-506	122-632	134-39	N/S	134-39	125-208	.031
0-7986	4160	600	37-1536	37-933	6-506	125-652	134-39	N/S	134-39	125-208	.031
0-7987	4160	600	37-1536	37-933	6-506	122-612	134-39	N/S	134-39	125-208	.031
0-8001	4360	450	37-1540	N/A	N/S	124-215	124-550	N/R	N/R	N/S	.028
0-8002	4360	450	37-1540	N/A	N/S	124-215	124-550	N/R	N/R	N/S	.028
0-8003	4360	450	37-1540	N/A	N/S	124-235	124-550	N/R	N/R	N/S	.028
0-8004	4160	600	37-1536	37-933	6-506	122-632	134-39	N/S	134-39	125-208	.031
0-8005	4160	600	37-1536	37-933	6-506	122-622	134-39	N/S	134-39	125-208	.031
0-8006	4160	600	37-1536	37-933	6-506	122-622	134-39	N/S	134-39	125-208	.031

Secondary Nozzle Size or Spring Color	Primary Bowl Gasket†	Primary Metering Block Gasket†	Secondary Bowl Gasket†	Secondary Metering Block Gasket†	Secondary Metering Plate Gasket†	Primary Fuel Bowl	Secondary Fuel Bowl	Throttle Body & Shaft Assembly	Venturi Diameter Primary	Venturi Diameter Secondary	Throttle Bore Diameter Primary	Throttle Bore Diameter Secondary
Black	108-83-2	108-91-2	108-90-2	108-90-2	108-27-2	34R8242AQ	134-105	N/S	1-1/4	1-5/16	1-9/16	1-9/16
Plain	108-83-2	108-91-2	108-90-2	108-90-2	108-27-2	N/S	N/S	N/S	1-3/16	1-1/4	1-1/2	1-1/2
Plain	108-83-2	108-91-2	108-90-2	108-90-2	108-27-2	N/S	N/S	N/S	1-3/16	1-1/4	1-1/2	1-1/2
Black	108-83-2	108-91-2	108-90-2	108-90-2	108-27-2	134-101	134-105	N/S	1-1/4	1-5/16	1-9/16	1-9/16
Black	108-83-2	108-91-2	108-90-2	108-90-2	108-27-2	134-101	134-105	N/S	1-1/4	1-5/16	1-9/16	1-9/16
Black	108-83-2	108-91-2	108-90-2	108-90-2	108-27-2	34R8242AQ	134-105	N/S	1-1/4	1-5/16	1-9/16	1-9/16
.037	108-92-2	108-91-2	108-92-2	108-91-2	N/R	134-110	34R7201A	N/S	1-5/32	1-3/8	1-3/8	2
Black	108-92-2	108-91-2	108-90-2	108-90-2	108-27-2	134-110	34R7960A	N/S	1-5/32	1-3/8	1-3/8	2
Plain	108-92-2	108-91-2	108-90-2	108-90-2	108-27-2	N/S	34R7960A	N/S	1-5/32	1-3/8	1-3/8	2
Plain	108-92-2	108-91-2	108-90-2	108-90-2	108-27-2	N/S	34R7960A	N/S	1-5/32	1-3/8	1-3/8	2
Plain	108-92-2	108-91-2	108-90-2	108-90-2	108-27-2	N/S	34R7960A	N/S	1-5/32	1-3/8	1-3/8	2
Plain	108-92-2	108-91-2	108-90-2	108-90-2	108-27-2	N/S	34R7960A	N/S	1-5/32	1-3/8	1-3/8	2
Plain	108-92-2	108-91-2	108-90-2	108-90-2	108-27-2	N/S	34R7960A	N/S	1-5/32	1-3/8	1-3/8	2
Black	108-83-2	108-91-2	108-90-2	108-90-2	108-27-2	N/S	134-105	N/S	1-1/4	1-5/16	1-9/16	1-9/16
Black	108-92-2	108-91-2	108-90-2	108-90-2	108-27-2	N/S	134-102	N/S	1-1/4	1-9/16	1-1/2	1-3/4
Purple	108-83-2	108-91-2	108-90-2	108-90-2	108-27-2	N/S	N/S	N/S	1-1/4	1-5/16	1-9/16	1-9/16
.037	108-92-2	108-91-2	108-83-2	108-91-2	N/R	134-110	34R7201A	N/S	1-5/32	1-3/8	1-3/8	2
Purple	108-83-2	108-91-2	108-90-2	108-90-2	108-27-2	N/S	N/S	N/S	1-1/4	1-5/16	1-9/16	1-9/16
.035	108-93	108-94	108-93	108-94	N/R	134-108	134-112	N/S	1-13/16	1-13/16	2	2
.035	108-93	108-94	108-93	108-94	N/R	134-108	134-112	N/S	1-13/16	1-13/16	2	2
N/R	N/R	N/R	N/R	N/R	N/R	N/R	N/R	N/R	1-1/25	1-1/16	1-7/25	1-7/16
N/R	N/R	N/R	N/R	N/R	N/R	N/R	N/R	N/R	1-1/33	1-1/16	1-1/4	1-7/25
Black	108-92-2	108-91-2	108-90-2	108-90-2	108-27-2	134-110	34R7960A	N/S	1-13/64	1-13/32	1-3/8	2
Black	108-92-2	108-91-2	108-90-2	108-90-2	108-27-2	134-110	34R7960A	N/S	1-13/64	1-13/32	1-3/8	2
.025	108-83-2	108-89-2	108-83-2	108-89-2	N/R	N/S	N/S	N/S	1-1/16	1-1/16	1-7/16	1-7/16
.025	108-83-2	108-89-2	108-83-2	108-89-2	N/R	N/S	N/S	N/S	1-1/16	1-1/16	1-7/16	1-7/16
Purple	108-83-2	108-91-2	108-90-2	108-90-2	108-27-2	N/S	N/S	N/S	1-1/4	1-5/16	1-9/16	1-9/16
N/R	108-83-2	108-89-2	N/R	N/R	N/R	134-103	N/R	12R11070A	1-3/16	N/R	1-1/2	N/R
N/R	N/S	N/R	N/R	N/R	N/R	N/R	N/R	N/S	1-1/16	1-3/16	1-3/8	1-7/16
N/R	N/S	N/R	N/R	N/R	N/R	N/R	N/R	N/S	1-1/16	1-3/16	1-3/8	1-7/16
N/R	N/S	N/R	N/R	N/R	N/R	N/R	N/R	N/S	1-1/16	1-3/16	1-3/8	1-7/16
Plain	108-83-2	108-91-2	108-90-2	108-90-2	108-27-2	N/S	N/S	N/S	1-1/4	1-5/16	1-9/16	1-9/16
Plain	108-92-2	108-91-2	108-90-2	108-90-2	108-27-2	N/S	34R7960A	N/S	1-13/32	1-13/64	1-3/8	2
N/R	N/S	N/R	N/R	N/R	N/R	N/R	N/R	N/S	1-1/16	1-3/16	1-3/8	1-7/16
N/R	N/S	N/R	N/R	N/R	N/R	N/R	N/R	N/S	1-1/16	1-3/16	1-3/8	1-7/16
N/R	N/S	N/R	N/R	N/R	N/R	N/R	N/R	N/S	1-1/16	1-3/16	1-3/8	1-7/16
Black	108-83-2	108-91-2	108-90-2	108-90-2	108-27-2	134-101	134-105	N/S	1-1/4	1-5/16	1-9/16	1-9/16
Black	108-83-2	108-91-2	108-90-2	108-90-2	108-27-2	N/S	134-105	N/S	1-1/4	1-5/16	1-9/16	1-9/16
Black	108-83-2	108-91-2	108-90-2	108-90-2	108-27-2	N/S	134-105	N/S	1-1/4	1-5/16	1-9/16	1-9/16
N/R	N/S	N/R	N/R	N/R	N/R	N/R	N/R	N/S	1-1/16	1-3/16	1-3/8	1-7/16
N/R	N/S	N/R	N/R	N/R	N/R	N/R	N/R	N/S	1-1/16	1-3/16	1-3/8	1-7/16
Black	108-83-2	108-91-2	108-90-2	108-90-2	108-27-2	N/S	134-105	N/S	1-1/4	1-5/16	1-9/16	1-9/16
Black	108-83-2	108-91-2	108-90-2	108-90-2	108-27-2	N/S	134-105	N/S	1-1/4	1-5/16	1-9/16	1-9/16
Black	108-83-2	108-91-2	108-90-2	108-90-2	108-27-2	N/S	134-105	N/S	1-1/4	1-5/16	1-9/16	1-9/16

APPENDIX - HOLLEY CARBURETOR NUMERICAL LISTINGS

Carburetor Part No.	Carb. Model No.	CFM	Renew Kit	Trick Kit	Primary & Secondary Needle & Seat	Primary Main Jet	Secondary Main Jet or Plate	Primary Metering Block	Secondary Metering Block	Primary Power Valve	Primary Discharge Nozzle Size
0-8007	4160	390	37-720	37-933	6-506	122-51	34R9716-59	34R8909AS	34R9716-59	125-65	.025
0-8059	4175	650	37-1537	37-933	(16,17)	122-582	134-21	N/S	134-21	125-206	.037
0-8059-1	4175	650	37-1537	37-933	(16,17)	122-582	N/S	N/S	N/S	125-211	.025
0-8060	4175	650	37-1537	37-933	(16,17)	122-582	134-21	N/S	134-21	125-206	.037
0-8060-1	4175	650	37-1537	37-933	(16,17)	122-582	N/S	N/S	N/S	N/S	.025
0-8082	4500	1050	37-1539	37-933	6-504	122-84	122-84	N/S	N/S	125-65	.035
0-8082-1	4500	1050	37-1539	37-933	6-504	122-88	122-88	N/S	N/S	125-65	.035
0-8082-2	4500	1050	37-1539	37-933	6-518-2	122-84	122-84	34R12013A	34R12013A	125-65 (15)	.035
0-8149	4360	450	37-1540	N/A	N/S	124-231	124-550	N/R	N/R	N/S	.028
0-8149-1	4360	450	37-1540	N/A	N/S	124-215	124-550	N/R	N/R	N/S	.028
0-8156	4150	750	37-485	37-933	6-504	122-70	122-83	134-155	N/S	125-65	.028
0-8158	4360	450	37-1540	N/A	N/S	124-219	124-550	N/R	N/S	N/S	.028
0-8162	4150	850	37-485	37-933	6-504	122-80	122-80	34R8558AS	N/S	125-65	.031
0-8181	4160	600	37-1536	37-933	6-504	122-80	122-80	N/S	134-39	125-65 (15)	.031
0-8203	4360	450	37-1540	N/A	N/S	124-211	124-550	N/R	N/R	N/S	.028
0-8204	4360	450	37-1540	N/A	N/S	124-215	124-550	N/R	N/R	N/S	.028
0-8206	4360	450	37-1540	N/A	N/S	124-203	124-550	N/R	N/R	N/S	.028
0-8207	4160	600	N/A	N/A	6-506	122-622	134-39	N/S	N/S	125-85	.031
0-8276	4175	650	37-1537	37-933	(16,17)	122-572	134-21	N/S	N/S	125-85	.025
0-8302	4175	650	37-1537	37-933	(16,17)	122-582	134-21	N/S	N/S	125-85	.025
0-8479	4360	450	37-1540	N/A	N/S	124-219	124-589	N/R	N/R	N/S	.028
0-8516	4360	450	37-1540	N/A	N/S	124-167	124-423	N/R	N/R	N/S	.028
0-8517	4360	450	37-1540	N/A	N/S	124-203	124-524	N/R	N/R	N/S	.028
0-8546	4175	650	37-1537	37-933	(16,17)	122-582	134-21	N/S	134-21	125-85	.025
0-8642	4360	450	37-1540	N/A	N/S	124-215	124-500	N/R	N/R	N/S	.028
0-8677	4360	450	37-1540	N/A	N/S	124-219	124-524	N/R	N/R	N/S	.028
0-8679	4175	650	37-1537	37-933	(16,17)	122-592	34R9716-27	N/S	34R9716-27	125-85	.025
0-8700	4175	650	37-1537	37-933	(16,17)	122-582	134-21	N/S	134-21	125-85	.025
0-8771	4360	450	37-1540	N/A	N/S	124-207	124-537	N/R	N/R	N/S	.028
0-8804	4150	830	37-485	37-933	6-504	122-80	122-80	N/S	N/S	125-65 (B)	.028
0-8874	4360	450	37-1540	N/A	N/S	124-219	124-589	N/R	N/R	N/S	.028
0-8875	4360	450	3-1160	N/A	N/S	124-231	124-576	N/R	N/R	N/S	.028
0-8876	4360	450	N/A	N/A	N/S	124-231	124-550	N/R	N/R	N/S	.028
0-8877	4360	450	3-1160	N/A	N/S	124-231	124-550	N/R	N/R	N/S	.028
0-8879	4175	650	37-1537	37-933	(16,17)	122-592	134-21	N/S	134-21	125-65	.025
0-8896	4500	1050	37-1539	37-933	6-504	122-88	122-88	34R9565AS	34R9565AS	N/R	.035
0-8896-1	4500	1050	37-1539	37-933	6-518-2	122-88	122-88	34R11972-1A	34R11972-1A	N/R	.035
0-8914	4360	450	37-1540	N/A	N/S	124-207	124-537	N/R	N/R	N/S	.028
0-8958	4360	450	37-1540	N/A	N/S	124-195	124-550	N/R	N/R	N/S	.028
0-9002	4160	600	37-1536	37-933	6-506	122-632	134-37	N/S	134-37	125-208	.031
0-9040	4160	600	37-119	37-933	N/S	122-661	N/S	N/S	N/S	125-211	.031
0-9088	4360	450	N/A	N/A	N/S	124-215	124-550	N/R	N/R	N/S	.028
0-9105	4360	450	3-1160	N/A	N/S	124-195	124-550	N/R	N/R	N/S	.028
0-9112	4360	450	37-1540	N/A	N/S	124-211	124-563	N/R	N/R	N/S	.028
0-9162	4360	450	37-1540	N/A	N/S	124-203	124-537	N/R	N/R	N/S	.028
0-9185	4360	450	37-1540	N/A	N/S	124-191	124-550	N/R	N/R	N/S	.028
0-9188	4150	780	37-1539	37-933	6-504	122-72	122-76	N/S	N/S	(12,21)	.025
0-9192	4360	450	37-1540	N/A	N/S	124-231	124-550	N/R	N/R	N/S	.028
0-9193	4360	450	37-1540	N/A	N/S	124-211	124-589	N/R	N/R	N/S	.028

© **HOLLEY CARBURETORS**
VISIT <WWW.HOLLEY.COM> FOR CURRENT VERSIONS

Secondary Nozzle Size or Spring Color	Primary Bowl Gasket†	Primary Metering Block Gasket†	Secondary Bowl Gasket†	Secondary Metering Block Gasket†	Secondary Metering Plate Gasket†	Primary Fuel Bowl	Secondary Fuel Bowl	Throttle Body & Shaft Assembly	Venturi Diameter Primary	Venturi Diameter Secondary	Throttle Bore Diameter Primary	Throttle Bore Diameter Secondary
Plain	108-83-2	108-91-2	108-90-2	108-90-2	108-27-2	34R8242AQ	134-105	12R7800-3AM	1-1/16	1-1/16	1-7/16	1-7/16
Black	108-92-2	108-91-2	108-90-2	108-90-2	108-27-2	134-110	34R7960A	N/S	1-13/64	1-13/32	1-3/8	2
Black	108-92-2	108-91-2	108-90-2	108-90-2	108-27-2	134-110	34R7960A	N/S	1-13/64	1-13/32	1-3/8	2
Black	108-92-2	108-91-2	108-90-2	108-90-2	108-27-2	134-110	34R7960A	N/S	1-13/64	1-13/32	1-3/8	2
Black	108-92-2	108-91-2	108-90-2	108-90-2	108-27-2	134-110	34R7960A	N/S	1-13/64	1-13/32	1-3/8	2
.035	108-93	108-94	108-93	108-94	N/R	134-108	134-112	N/S	1-11/16	1-11/16	2	2
.035	108-93	108-94	108-93	108-94	N/R	134-108	134-112	N/S	1-11/16	1-11/16	2	2
N/R	N/S	N/R	N/R	N/R	N/R	N/R	N/R	N/S	1-1/16	1-3/16	1-3/8	1-7/16
N/R	N/S	N/R	N/R	N/R	N/R	N/R	N/R	N/S	1-1/16	1-3/16	1-3/8	1-7/16
.031	108-83-2	108-89-2	108-83-2	108-89-2	N/R	134-103	134-104	12R8039-3AM	1-3/8	1-3/8	1-11/16	1-11/16
N/R	N/S	N/R	N/R	N/R	N/R	N/R	N/R	N/S	1-1/16	1-3/16	1-3/8	1-7/16
.031	108-83-2	108-89-2	108-83-2	108-89-2	N/R	134-103	134-104	12R8053-3AM	1-9/16	1-9/16	1-3/4	1-3/4
.031	108-83-2	108-89-2	108-83-2	108-89-2	N/R	34R8242AQ	134-105	N/S	1-9/16	1-9/16	1-3/4	1-3/4
N/R	N/S	N/R	N/R	N/R	N/R	N/R	N/R	N/S	1-1/16	1-3/16	1-3/8	1-7/16
N/R	N/S	N/R	N/R	N/R	N/R	N/R	N/R	N/S	1-1/16	1-3/16	1-3/8	1-7/16
N/R	N/S	N/R	N/R	N/R	N/R	N/R	N/R	N/S	1-1/16	1-3/16	1-3/8	1-7/16
Plain	108-83-2	108-91-2	108-90-2	108-90-2	108-27-2	N/S	N/S	N/S	1-1/4	1-5/16	1-9/16	1-9/16
Black	108-92-2	108-91-2	108-90-2	108-90-2	108-27-2	N/S	N/S	N/S	1-13/64	1-13/32	1-3/8	2
Black	108-92-2	108-91-2	108-90-2	108-90-2	108-27-2	N/S	N/S	N/S	1-13/64	1-13/32	1-3/8	2
N/R	N/S	N/R	N/R	N/R	N/R	N/R	N/R	N/S	1-1/16	1-3/16	1-3/8	1-7/16
N/R	N/S	N/R	N/R	N/R	N/R	N/R	N/R	N/S	1-1/16	1-3/16	1-3/8	1-7/16
Black	108-92-2	108-91-2	108-90-2	108-90-2	108-27-2	134-110	34R7960A	N/S	1-13/64	1-13/32	1-3/8	2
N/R	N/S	N/R	N/R	N/R	N/R	N/R	N/R	N/S	1-1/16	1-3/16	1-3/8	1-7/16
N/R	N/S	N/R	N/R	N/R	N/R	N/R	N/R	N/S	1-1/16	1-3/16	1-3/8	1-7/16
Plain	108-92-2	108-91-2	108-90-2	108-90-2	108-27-2	134-110	34R7960A	N/S	1-13/64	1-13/32	1-3/8	2
Black	108-92-2	108-91-2	108-90-2	108-90-2	108-27-2	134-110	34R7960A	N/S	1-13/64	1-13/32	1-3/8	2
N/R	N/S	N/R	N/R	N/R	N/R	N/R	N/R	N/S	1-1/16	1-3/16	1-3/8	1-7/16
.028	108-83-2	108-89-2	108-83-2	108-89-2	N/R	134-103	134-104	12R8434-3AM	1-9/16	1-9/16	1-11/16	1-11/16
N/R	N/S	N/R	N/R	N/R	N/R	N/R	N/R	N/S	1-1/16	1-3/16	1-3/8	1-7/16
N/R	N/S	N/R	N/R	N/R	N/R	N/R	N/R	N/S	1-1/16	1-3/16	1-3/8	1-7/16
N/R	N/S	N/R	N/R	N/R	N/R	N/R	N/R	N/S	1-1/16	1-3/16	1-3/8	1-7/16
Black	108-92-2	108-91-2	108-90-2	108-90-2	108-27-2	134-110	34R7960A	N/S	1-13/64	1-13/32	1-3/8	2
.035	108-95	108-96	108-95	108-96	N/R	134-108	134-112	N/S	1-11/16	1-11/16	2	2
.037	108-95	108-96	108-95	108-96	N/R	134-108	134-112	N/S	1-11/16	1-11/16	2	2
N/R	N/S	N/R	N/R	N/R	N/R	N/R	N/R	N/S	1-1/16	1-3/16	1-3/8	1-7/16
N/R	N/S	N/R	N/R	N/R	N/R	N/R	N/R	N/S	1-1/16	1-3/16	1-3/8	1-7/16
Black	108-83-2	108-91-2	108-90-2	108-90-2	108-27-2	34R8242AQ	134-105	N/S	1-1/4	1-5/16	1-9/16	1-9/16
Plain	108-83-2	108-91-2	108-90-2	108-90-2	108-27-2	N/S	134-105	N/S	1-1/4	1-5/16	1-9/16	1-9/16
N/R	N/S	N/R	N/R	N/R	N/R	N/R	N/R	N/S	1-1/16	1-3/16	1-3/8	1-7/16
N/R	N/S	N/R	N/R	N/R	N/R	N/R	N/R	N/S	1-1/16	1-3/16	1-3/8	1-7/16
N/R	N/S	N/R	N/R	N/R	N/R	N/R	N/R	N/S	1-1/16	1-3/16	1-3/8	1-7/16
N/R	N/S	N/R	N/R	N/R	N/R	N/R	N/R	N/S	1-1/16	1-3/16	1-3/8	1-7/16
Plain	108-83-2	108-89-2	108-83-2	108-89-2	N/R	N/S	N/S	N/S	1-3/8	1-7/16	1-11/16	1-11/16
N/R	N/S	N/R	N/R	N/R	N/R	N/R	N/R	N/S	1-1/16	1-3/16	1-3/8	1-7/16
N/R	N/S	N/R	N/R	N/R	N/R	N/R	N/R	N/S	1-1/16	1-3/16	1-3/8	1-7/16

APPENDIX - HOLLEY CARBURETOR NUMERICAL LISTINGS

Carburetor Part No.	Model No.	Carb. CFM	Renew Kit	Trick Kit	Secondary Needle	Primary & Main Jet & Seat	Primary Main Jet or Plate	Secondary Metering Block	Primary Metering Block	Secondary Primary PowerValve	Primary Discharge Nozzle Size
0-9210	4160	600	37-1536	37-933	6-506	122-612	134-39	N/S	N/S	125-208	.031
0-9219	4160	600	37-1536	37-933	6-506	122-632	134-39	N/S	134-39	125-208	.031
0-9228	5200	280	N/A	N/A	N/S	124-163	124-231	N/R	N/R	N/S	.023
0-9254	4160	600	37-1536	37-933	6-506	122-622	134-39	N/S	N/S	125-211	.031
0-9375	4500	1050	37-1539	37-933	6-504	122-92	122-92	34R9565AS	34R9565AS	N/R	.035
0-9375-1	4500	1050	37-1539	37-933	6-518-2	122-88	122-88	34R11972-2A	34R11972-2A	N/R	.035
0-9377	4500	1150	37-1539	37-933	6-504	122-94	122-94	N/S	N/S	N/R	.035
0-9377-1	4500	1150	37-1539	37-933	6-518-2	122-92	122-92	34R11972-3A	34R9716-27	N/R	.035
0-9379	4150	750	37-485	37-933	6-504	122-68	122-81	134-155	N/S	125-65	.028
0-9380	4150	850	37-485	37-933	6-504	122-78	122-78	34R8558AS	N/S	125-65	(15) .031
0-9381	4150	830	37-485	37-933	6-504	122-78	122-78	34R8558AS	N/S	125-65	(15) .028
0-9429	5200	280	N/A	N/A	N/S	124-183	124-231	N/R	N/R	N/S	.023
0-9441	5200	280	N/A	N/A	N/S	124-163	124-231	N/R	N/R	N/S	.023
0-9444	5200	280	N/A	N/A	N/S	124-163	124-231	N/R	N/R	N/S	.023
0-9446	5200	280	N/A	N/A	N/S	124-163	124-231	N/R	N/R	N/S	.023
0-9545	5200	280	N/A	N/A	N/S	124-183	124-231	N/R	N/R	N/S	.023
0-9626	4160	600	3-1415	N/A	6-506	122-612	134-39	N/S	N/S	125-206	.031
0-9644	6520	280	N/A	N/A	N/A	124-179	124-283	N/R	N/R	N/A	.020
0-9645	4150	750	37-1539	37-933	6-515-2	122-80	122-80	34R9929AS	34R9936AS	125-165	(15) .045
0-9646	4150	850	37-1539	37-933	6-515-2	122-92	122-92	34R9929AS	34R9936AS	125-165	(15) .045
0-9647	2300	500	37-1536	37-933	6-515-2	122-81	N/R	34R9925AS	N/R	125-145	.040
0-9655	6520	280	N/A	N/A	N/A	124-195	124-299	N/R	N/R	N/A	.020
0-9659	6520	280	N/A	N/A	N/A	124-131	124-267	N/R	N/R	N/A	.020
0-9678	4360	450	3-1160	N/A	N/S	124-211	124-550	N/R	N/R	N/S	.028
0-9681	5200	280	N/A	N/A	N/A	124-171	124-215	N/R	N/R	N/A	.023
0-9682	6520	280	N/A	N/A	N/A	124-219	124-283	N/R	N/R	N/A	.020
0-9688	5200	280	N/A	N/A	N/S	124-163	124-251	N/R	N/R	N/S	.023
0-9689	5200	280	N/A	N/A	N/S	124-159	124-251	N/R	N/R	N/S	.023
0-9694	4360	450	37-1540	N/A	N/S	124-171	124-485	N/R	N/R	N/S	.028
0-9767	5200	280	N/A	N/A	N/S	124-179	124-259	N/R	N/R	N/S	.023
0-9776	4160	450	37-1536	37-933	6-506	122-582	34R9716-6	N/S	34R9716-6	125-85	.031
0-9777	4360	450	37-1540	N/A	N/S	124-255	124-550	N/R	N/R	N/S	.028
0-9781	5200	280	N/A	N/A	N/S	124-159	124-251	N/R	N/R	N/S	.023
0-9810	6520	280	N/A	N/A	N/A	124-195	124-299	N/R	N/R	N/A	.020
0-9811	6520	280	N/A	N/A	N/A	124-155	124-271	N/R	N/R	N/A	.020
0-9834	4160	600	37-720	37-933	6-506	122-642	134-39	N/S	134-39	125-65	.031
0-9834-1	4160	600	37-720	37-933	6-506	122-661	134-39	N/S	134-39	125-65	.031
0-9834-2	4160	600	37-720	37-933	6-506	122-68	134-39	N/S	134-39	125-65	.031
0-9834-3	4160	600	37-720	37-933	6-506	122-68	134-39	N/S	134-39	125-65	.031
0-9864	5200	280	N/A	N/A	N/S	124-159	124-219	N/R	N/R	N/S	.023
0-9875	4360	450	N/A	N/A	N/S	124-199	124-576	N/R	N/R	N/S	.028
0-9895	4175	650	37-1537	37-933	(16,17)	122-592	134-21	N/S	134-21	125-206	.037
0-9896	6510	280	N/A	N/A	N/A	124-104	124-271	N/R	N/R	N/A	.020
0-9899	5200	280	N/A	N/A	N/S	124-147	124-231	N/R	N/R	N/S	.023
0-9923	4175	650	37-1537	37-933	(16,17)	122-542	N/S	N/S	N/S	125-211	.025
0-9925	5200	280	N/A	N/A	N/S	124-147	124-251	N/R	N/R	N/S	.023
0-9931	4360	450	37-1540	N/A	N/S	124-239	124-550	N/R	N/R	N/S	.028
0-9932	5200	280	N/A	N/A	N/S	124-159	124-219	N/R	N/R	N/S	.023
0-9935	4360	450	37-1540	N/A	N/S	124-207	124-589	N/R	N/R	N/S	.028

© **HOLLEY CARBURETORS**
VISIT <WWW.HOLLEY.COM> FOR CURRENT VERSIONS

Secondary Nozzle Size or Spring Color	Primary Bowl Gasket†	Primary Metering Block Gasket†	Secondary Bowl Gasket†	Secondary Metering Block Gasket†	Secondary Metering Plate Gasket†	Primary Fuel Bowl	Secondary Fuel Bowl	Throttle Body & Shaft Assembly	Venturi Diameter Primary	Venturi Diameter Secondary	Throttle Bore Diameter Primary	Throttle Bore Diameter Secondary
Black	108-83-2	108-91-2	108-90-2	108-90-2	108-27-2	N/S	N/S	N/S	1-1/4	1-5/16	1-9/16	1-9/16
Black	108-83-2	108-91-2	108-90-2	108-90-2	108-27-2	N/S	134-105	N/S	1-1/4	1-5/16	1-9/16	1-9/16
N/R	N/R	N/R	N/R	N/R	N/R	N/R	N/R	N/R	1-1/25	1-1/16	1-7/25	1-7/16
Black	108-83-2	108-91-2	108-90-2	108-90-2	108-27-2	N/S	N/S	N/S	1-1/4	1-5/16	1-9/16	1-9/16
.035	108-95	108-96	108-95	108-96	N/R	134-108	134-112	N/S	1-11/16	1-11/16	2	2
.037	108-95	108-96	108-95	108-96	N/R	134-108	134-112	N/S	1-11/16	1-11/16	2	2
.035	108-95	108-96	108-95	108-96	N/R	134-108	134-112	N/S	1-13/16	1-13/16	2	2
.037	108-95	108-96	108-95	108-96	N/R	134-108	134-112	N/S	1-13/16	1-13/16	2	2
.031	108-83-2	108-89-2	108-83-2	108-89-2	N/R	134-103	134-104	12R8039-3AM	1-3/8	1-3/8	1-11/16	1-11/16
.031	108-83-2	108-89-2	108-83-2	108-89-2	N/R	134-103	134-104	12R8053-3AM	1-9/16	1-9/16	1-3/4	1-3/4
.028	108-83-2	108-89-2	108-83-2	108-89-2	N/R	134-103	134-104	12R8434-3AM	1-9/16	1-9/16	1-11/16	1-11/16
N/R	N/R	N/R	N/R	N/R	N/R	N/R	N/R	N/R	1-1/25	1-1/16	1-7/25	1-7/16
N/R	N/R	N/R	N/R	N/R	N/R	N/R	N/R	N/R	1-1/25	1-1/16	1-7/25	1-7/16
N/R	N/R	N/R	N/R	N/R	N/R	N/R	N/R	N/R	1-1/25	1-1/16	1-7/25	1-7/16
N/R	N/R	N/R	N/R	N/R	N/R	N/R	N/R	N/R	1-1/25	1-1/16	1-7/25	1-7/16
Black	108-83-2	108-91-2	108-90-2	108-90-2	108-27-2	N/S	N/S	N/S	1-1/4	1-5/16	1-9/16	1-9/16
N/R	N/R	N/R	N/R	N/R	N/R	N/R	N/R	N/R	1-1/25	1-1/16	1-7/25	1-7/16
.045	108-83-2	108-89-2	108-83-2	108-89-2	N/R	134-103	134-104	12R9182A	1-3/8	1-3/8	1-11/16	1-11/16
.045	108-83-2	108-89-2	108-83-2	108-89-2	N/R	134-103	134-104	N/S	1-9/16	1-9/16	1-3/4	1-3/4
N/R	108-83-2	108-89-2	N/R	N/R	N/R	134-103	N/S	112-2	1-3/8	N/R	1-11/16	N/R
N/R	N/R	N/R	N/R	N/R	N/R	N/R	N/R	N/R	1-1/25	1-1/16	1-7/25	1-7/16
N/R	N/S	N/R	N/R	N/R	N/R	N/R	N/R	N/S	1-1/16	1-3/16	1-3/8	1-7/16
N/R	N/R	N/R	N/R	N/R	N/R	N/R	N/R	N/R	1-1/25	1-1/16	1-7/25	1-7/16
N/R	N/R	N/R	N/R	N/R	N/R	N/R	N/R	N/R	1-1/25	1-1/16	1-7/25	1-7/16
N/R	N/R	N/R	N/R	N/R	N/R	N/R	N/R	N/R	1-1/25	1-1/16	1-7/25	1-7/16
N/R	N/S	N/R	N/R	N/R	N/R	N/R	N/R	N/R	1-1/25	1-1/16	1-3/8	1-7/16
N/R	N/R	N/R	N/R	N/R	N/R	N/R	N/R	N/R	1-1/25	1-1/16	1-7/25	1-7/16
N/R	108-83-2	108-89-2	108-90-2	108-90-2	108-27-2	134-101	134-105	12R9384A	1-3/32	1-3/32	1-1/2	1-1/2
N/R	N/S	N/R	N/R	N/R	N/R	N/R	N/R	N/R	1-1/16	1-3/16	1-3/8	1-7/16
N/R	N/R	N/R	N/R	N/R	N/R	N/R	N/R	N/R	1-1/25	1-1/16	1-7/25	1-7/16
N/R	N/R	N/R	N/R	N/R	N/R	N/R	N/R	N/R	1-1/25	1-1/16	1-7/25	1-7/16
Black	108-83-2	108-91-2	108-90-2	108-90-2	108-27-2	34R8242AQ	134-105	N/S	1-1/4	1-5/16	1-9/16	1-9/16
Black	108-83-2	108-91-2	108-90-2	108-90-2	108-27-2	34R8242AQ	134-105	N/S	1-1/4	1-5/16	1-9/16	1-9/16
Black	108-83-2	108-91-2	108-90-2	108-90-2	108-27-2	34R8242AQ	134-105	N/S	1-1/4	1-5/16	1-9/16	1-9/16
Black	108-83-2	108-91-2	108-90-2	108-90-2	108-27-2	34R8242AQ	134-105	N/S	1-1/4	1-5/16	1-9/16	1-9/16
N/R	N/S	N/R	N/R	N/R	N/R	N/R	N/R	N/R	1-1/25	1-1/16	1-7/25	1-7/16
N/R	N/S	N/R	N/R	N/R	N/R	N/R	N/R	N/R	1-1/16	1-1/16	1-3/8	1-7/16
Black	108-92-2	108-91-2	108-90-2	108-90-2	108-27-2	134-110	34R7960A	12R9482A	1-13/64	1-13/32	1-3/8	2
N/R	N/R	N/R	N/R	N/R	N/R	N/R	N/R	N/R	1-1/25	1-1/16	1-7/25	1-7/16
Black	108-92-2	108-91-2	108-90-2	108-90-2	108-27-2	N/S	N/S	N/S	1-13/64	1-13/32	1-3/8	2
N/R	N/R	N/R	N/R	N/R	N/R	N/R	N/R	N/R	1-1/25	1-1/16	1-7/25	1-7/16
N/R	N/S	N/R	N/R	N/R	N/R	N/R	N/R	N/S	1-1/16	1-3/16	1-3/8	1-7/16
N/R	N/R	N/R	N/R	N/R	N/R	N/R	N/R	N/R	1-1/25	1-1/16	1-7/25	1-7/16
N/R	N/S	N/R	N/R	N/R	N/R	N/R	N/R	N/S	1-1/16	1-3/16	1-3/8	1-7/16

APPENDIX - HOLLEY CARBURETOR NUMERICAL LISTINGS

Carburetor Part No.	Carb. Model No.	CFM	Renew Kit	Trick Kit	Primary & Secondary Needle & Seat	Primary Main Jet	Secondary Main Jet or Plate	Primary Metering Block	Secondary Metering Block	Primary Power Valve	Primary Discharge Nozzle Size
0-9948	4175	650	37-1537	37-933	(16,17)	122-563	N/S	N/S	N/S	125-211	.025
0-9973	4360	450	37-1540	N/A	N/S	124-171	124-330	N/R	N/R	N/S	.028
0-9976	4175	650	37-1537	37-933	(16,17)	122-582	N/S	N/S	N/S	125-211	.025
0-50399	4160	650	703-28	N/A	6-511	122-73	N/S	N/S	N/R	125-65	.040
0-50399-1	4160	650	703-28	N/A	N/S	122-73	N/S	N/S	N/R	125-65	.040
0-80054	5200	280	N/A	N/A	N/S	124-231	124-247	N/R	N/R	N/S	.023
0-80055	5200	280	N/A	N/A	N/S	124-231	124-247	N/R	N/R	N/S	.023
0-80056	5200	280	N/A	N/A	N/S	124-231	124-247	N/R	N/R	N/S	.023
0-80057	5200	280	N/A	N/A	N/S	124-132	124-135	N/R	N/R	N/S	.023
0-80073	4175	650	N/A	N/A	(16,17)	122-642	N/S	N/S	N/S	125-213	.037
0-80086	4360	450	N/A	N/A	N/S	124-199	124-550	N/R	N/R	N/S	.028
0-80095	2305	500	37-1536	37-933	6-504	122-55	122-73	N/S	N/R	125-85	.035
0-80098	4180	600	37-1536	37-933	6-517	122-612	N/S	N/S	N/S	125-215	.028
0-80099	4180	600	37-1536	37-933	6-517	122-622	N/S	N/S	N/S	125-218	.028
0-80111	4180	600	37-1536	37-933	6-517	122-612	N/S	N/S	N/S	125-216	.028
0-80112	4180	600	37-1536	37-933	6-517	122-622	N/S	N/S	N/S	125-217	.028
0-80120	2305	350	37-1536	37-933	6-504	122-52	122-65	N/S	N/R	125-85	.035
0-80128	4175	650	37-1537	37-933	6-510	122-582	N/S	N/S	N/S	125-211	.031
0-80133	4180	600	37-1536	37-933	6-517	122-611	N/S	N/S	N/S	125-216	.028
0-80134	4180	600	37-1536	37-933	6-517	122-612	N/S	N/S	N/S	N/S	.028
0-80135	4180	600	37-1536	37-933	6-517	122-612	N/S	N/S	N/S	N/S	.028
0-80136	4180	600	37-1536	37-933	6-517	122-612	N/S	N/S	N/S	N/S	.028
0-80137	4180	600	37-1536	37-933	6-517	122-612	N/S	N/S	N/S	N/S	.028
0-80139	4175	650	37-1537	37-933	6-510	122-592	134-21	N/S	N/S	N/S	.037
0-80140	4175	650	N/A	N/A	6-510	122-642	N/S	N/S	N/S	125-213	.037
0-80145	4150	600	37-1539	37-933	6-504	122-68	122-70	N/S	N/S	125-65	.031
0-80155	4175	650	37-1537	37-933	6-510	122-632	134-21	N/S	N/S	N/S	.037
0-80163	4180	600	37-1536	37-933	6-517	122-622	N/S	N/S	N/S	N/S	.028
0-80164	4180	600	37-1536	37-933	6-517	122-612	N/S	N/S	N/S	N/S	.028
0-80165	4180	600	37-1536	37-933	6-517	122-612	N/S	N/S	N/S	N/S	.028
0-80166	4180	600	37-1536	37-933	6-517	122-612	N/S	N/S	N/S	N/S	.028
0-80169	4175	650	37-1537	37-933	6-510	122-543	34R5113-3	N/S	N/S	125-211	.025
0-80186	4500	750	37-1539	37-933	6-504	122-70	122-70	N/S	N/S	125-65 (15)	.028
0-80186-1	4500	750	37-1539	37-933	6-518-2	122-70	122-70	N/S	N/S	125-65 (15)	.028
0-80431	4160	550	37-119	37-933	6-506	122-60	134-9	N/S	N/S	125-65	.025
0-80432	4160	550	37-119	37-933	6-506	122-60	134-9	N/S	N/S	125-65	.025
0-80436	4150	850	37-1539	37-933	6-504	122-80	122-80	N/S	34R11698AQ	125-65 (22)	.040
0-80450	4160	600	37-1536	37-933	6-506	122-622	134-39	N/S	N/S	125-208	.031
0-80451	4160	600	37-1536	37-933	6-506	122-622	134-39	N/S	N/S	125-208	.031
0-80452	4160	600	37-1536	37-933	6-506	122-652	134-39	N/S	N/S	125-208	.031
0-80453	4160	600	37-1536	37-933	6-506	122-632	134-39	N/S	N/S	125-208	.031
0-80454	4160	600	37-1536	37-933	6-506	122-622	134-39	N/S	N/S	125-208	.031
0-80457	4160	600	37-119	37-933	6-506	122-69	134-39	134-128	134-39	125-65	.031
0-80457-1	4160	600	37-119	37-933	6-506	122-64	134-39	134-128	134-39	125-65	.031
0-80457-2	4160	600	37-119	37-933	6-506	122-64	134-39	134-128S	134-39	125-65	.031
0-80460	4160	600	37-1536	37-933	6-506	122-622	134-39	N/S	N/S	125-208	.031
0-80491	4175	650	37-1537	37-933	6-511	122-632	134-21	N/S	134-21	N/S	.037
0-80496	4150	950	37-1539	37-933	6-518-2	122-78	122-78	34R11845A	34R11845A	125-165 (both)	.031
0-80497	4150	950	37-1539	37-933	6-518-2	122-78	122-78	34R11845A	34R11845A	125-165 (both)	.031

HOLLEY CARBURETORS
VISIT <WWW.HOLLEY.COM> FOR CURRENT VERSIONS

Secondary Nozzle Size or Spring Color	Primary Bowl Gasket†	Primary Metering Block Gasket†	Secondary Bowl Gasket†	Secondary Metering Block Gasket†	Secondary Metering Plate Gasket†	Primary Fuel Bowl	Secondary Fuel Bowl	Throttle Body & Shaft Assembly	Venturi Diameter Primary	Venturi Diameter Secondary	Throttle Bore Diameter Primary	Throttle Bore Diameter Secondary
Black	108-92-2	108-91-2	108-90-2	108-90-2	108-27-2	N/S	N/S	N/S	1-13/64	1-13/32	1-3/8	2
N/R	N/S	N/R	N/R	N/R	N/R	N/R	N/R	N/S	1-1/16	1-3/16	1-3/8	1-7/16
Black	108-92-2	108-91-2	108-90-2	108-90-2	108-27-2	N/S	34R7960A	N/S	1-13/64	1-13/32	1-3/8	2
White	108-83-2	108-91-2	108-90-2	108-90-2	108-13-2	N/S	N/S	N/S	1-1/4	1-5/16	1-9/16	1-9/16
White	108-83-2	108-91-2	108-90-2	108-90-2	108-13-2	N/S	N/S	N/S	1-1/4	1-5/16	1-9/16	1-9/16
N/R	N/S	N/R	N/R	N/R	N/R	N/R	N/R	N/R	1-1/25	1-1/16	1-7/25	1-7/16
N/R	N/S	N/R	N/R	N/R	N/R	N/R	N/R	N/R	1-1/25	1-1/16	1-7/25	1-7/16
N/R	N/S	N/R	N/R	N/R	N/R	N/R	N/R	N/R	1-1/25	1-1/16	1-7/25	1-7/16
Black	34-202	108-91-2	108-90-2	108-90-2	108-27-2	N/S	N/S	N/S	1-13/64	1-13/32	1-3/8	2
N/R	N/S	N/R	N/R	N/R	N/R	N/R	N/R	N/R	1-1/16	1-3/16	1-3/8	1-7/16
.028	108-83-2	108-89-2	N/R	N/R	N/R	N/R	N/R	N/R	1-3/8	1-3/8	1-11/16	1-11/16
Purple	108-56-2	108-55-2	108-90-2	108-90-2	108-13-2	N/S	N/S	N/S	1-1/4	1-5/16	1-9/16	1-9/16
Purple	108-56-2	108-55-2	108-90-2	108-90-2	108-13-2	N/S	N/S	N/S	1-1/4	1-5/16	1-9/16	1-9/16
Orange	108-56-2	108-55-2	108-90-2	108-90-2	108-13-2	N/S	N/S	N/S	1-1/4	1-5/16	1-9/16	1-9/16
Orange	108-56-2	108-55-2	108-90-2	108-90-2	108-13-2	N/S	N/S	N/S	1-1/4	1-5/16	1-9/16	1-9/16
.028	108-83-2	108-89-2	N/R	N/R	N/R	N/S	N/S	N/S	1-3/16	1-3/16	1-11/16	1-11/16
White	108-92-2	108-91-2	108-90-2	108-90-2	108-27-2	N/S	N/S	N/S	1-13/32	1-13/64	1-3/8	2
Orange	108-56-2	108-55-2	108-90-2	108-90-2	108-13-2	N/S	N/S	N/S	1-1/4	1-5/16	1-9/16	1-9/16
Brown	108-56-2	108-55-2	108-90-2	108-90-2	108-13-2	N/S	N/S	N/S	1-1/4	1-5/16	1-9/16	1-9/16
Brown	108-56-2	108-55-2	108-90-2	108-90-2	108-13-2	N/S	N/S	N/S	1-1/4	1-5/16	1-9/16	1-9/16
Brown	108-56-2	108-55-2	108-90-2	108-90-2	108-13-2	N/S	N/S	N/S	1-1/4	1-5/16	1-9/16	1-9/16
Brown	108-56-2	108-55-2	108-90-2	108-90-2	108-13-2	N/S	N/S	N/S	1-1/4	1-5/16	1-9/16	1-9/16
Black	108-92-2	108-91-2	108-90-2	108-90-2	108-27-2	N/S	N/S	N/S	1-13/32	1-13/64	1-3/8	2
Black	34-202	108-91-2	108-90-2	108-90-2	108-27-2	N/S	N/S	N/S	1-13/32	1-13/64	1-3/8	2
Plain	108-83-2	108-91-2	108-83-2	108-89-2	N/R	N/S	34R11442	N/S	1-1/4	1-5/16	1-9/16	1-9/16
Black	108-92-2	108-91-2	108-90-2	108-90-2	108-27-2	N/S	N/S	N/S	1-13/32	1-13/64	1-3/8	2
Purple	108-56-2	108-55-2	108-90-2	108-90-2	108-13-2	N/S	N/S	N/S	1-1/4	1-5/16	1-9/16	1-9/16
Brown	108-56-2	108-55-2	108-90-2	108-90-2	108-13-2	N/S	N/S	N/S	1-1/4	1-5/16	1-9/16	1-9/16
Pink	108-56-2	108-55-2	108-90-2	108-90-2	108-13-2	N/S	N/S	N/S	1-1/4	1-5/16	1-9/16	1-9/16
Pink	108-56-2	108-55-2	108-90-2	108-90-2	108-13-2	N/S	N/S	N/S	1-1/4	1-5/16	1-9/16	1-9/16
Black	108-92-2	108-35-2	108-90-2	108-90-2	108-27-2	N/S	N/S	N/S	1-13/32	1-13/64	1-3/8	2
.035	108-93	108-94	108-93	108-94	N/R	134-108	134-112	N/S	1-11/16	1-11/16	2	2
.035	108-93	108-94	108-93	108-94	N/R	134-108	134-112	N/S	1-11/16	1-11/16	2	2
Plain	108-83-2	108-89-2	108-90-2	108-90-2	108-27-2	N/S	N/S	N/S	1-3/16	1-1/4	1-1/2	1-1/2
Plain	108-83-2	108-89-2	108-90-2	108-90-2	108-27-2	N/S	N/S	N/S	1-3/16	1-1/4	1-1/2	1-1/2
Pink	108-83-2	108-89-2	108-83-2	108-89-2	N/R	134-103	134-102	12R11052A	1-9/16	1-9/16	1-3/4	1-3/4
Black	108-83-2	108-89-2	108-90-2	108-90-2	N/R	N/S	N/S	N/S	1-1/4	1-5/16	1-9/16	1-9/16
Black	108-83-2	108-89-2	108-90-2	108-90-2	N/R	N/S	N/S	N/S	1-1/4	1-5/16	1-9/16	1-9/16
Black	108-83-2	108-89-2	108-90-2	108-90-2	N/R	N/S	N/S	N/S	1-1/4	1-5/16	1-9/16	1-9/16
Black	108-83-2	108-89-2	108-90-2	108-90-2	N/R	N/S	N/S	N/S	1-1/4	1-5/16	1-9/16	1-9/16
Black	108-83-2	108-89-2	108-90-2	108-90-2	N/R	134-101	134-105	N/S	1-1/4	1-5/16	1-9/16	1-9/16
Black	108-83-2	108-89-2	108-90-2	108-90-2	N/R	134-101	134-105	12R11240A	1-1/4	1-5/16	1-9/16	1-9/16
Black	108-83-2	108-89-2	108-90-2	108-90-2	N/R	134-101S	134-105S	12R11240A	1-1/4	1-5/16	1-9/16	1-9/16
Black	108-83-2	108-89-2	108-90-2	108-90-2	N/R	N/S	N/S	N/S	1-1/4	1-5/16	1-9/16	1-9/16
Black	108-92-2	108-91-2	108-90-2	108-90-2	108-27-2	N/S	34R7960A	N/S	1-13/32	1-13/64	1-3/8	2
.031	108-93	108-94	108-93	108-94	N/R	134-108	134-112	12R11194AS	1-3/8	1-3/8	1-3/4	1-3/4
Brown	108-93	108-94	108-93	108-94	N/R	134-108	34R11442	12R11199AS	1-3/8	1-3/8	1-3/4	1-3/4

APPENDIX - HOLLEY CARBURETOR NUMERICAL LISTINGS

Carburetor Part No.	Carb. Model No.	CFM	Renew Kit	Trick Kit	Primary & Secondary Needle & Seat	Primary Main Jet	Secondary Main Jet or Plate	Primary Metering Block	Secondary Metering Block	Primary Power Valve	Primary Discharge Nozzle Size
0-80498	4150	950	37-1539	37-933	6-519-2	122-144	122-144	34R11861A	34R11861A	125-155 (both)	.055
0-80507	4150	390	37-1539	37-933	6-504	122-65	122-65	34R11885A	34R11885A	125-35 (22)	.025
0-80508	4160	750	37-754	37-933	6-504	122-72	134-21	134-131	134-21	125-65	.025
0-80508-1	4160	750	37-754	37-933	6-504	122-72	134-21	134-131S	134-21	125-65	.025
0-80509	4150	830	37-1539	37-933	6-504	122-86	122-86	34R11895A	34R11895A	125-65 (15)	.028
0-80511	4150	830	37-1539	37-933	6-518-2	122-84	122-84	34R11899A	34R11899A	125-65 (15)	.028
0-80512	4150	1000	37-1539	37-933	6-518-2	122-84	122-84	34R11910A	34R11910A	125-65 (15)	.031
0-80513	4150	1000	37-1539	37-933	6-518-2	122-84	122-84	34R11910A	34R11910A	125-65 (15)	.031
0-80514	4150	1000	37-1539	37-933	6-518-2	122-84	122-88	34R11920A	34R11920A	125-65 (15)	.031
0-80519	4150	1000	37-1539	37-933	6-518-2	122-84	122-88	34R11920A	34R11920A	125-65 (15)	.031
0-80528	4150	750	37-1539	37-933	6-504	122-72	122-84	134-261	134-261	125-65	.031
0-80528-1	4150	750	37-1539	37-933	6-504	122-73	122-73	134-261	134-261	125-65 (15)	.031
0-80529	4150	750	37-1539	37-933	6-504	122-72	122-84	134-261	134-261	125-65	.031
0-80529-1	4150	750	37-1539	37-933	6-504						
0-80531	4150	850	37-1539	37-933	6-504	122-78	122-82	34R11702A	34R11699A	125-45 (22)	.040
0-80532	4500	1250	37-1539	37-933	6-518-2	122-97	122-97	34R11972-4A	34R11972-4A	N/R	.035
0-80533	4500	1250	37-1539	37-933	6-518-2	122-97	122-97	34R11972-5A	34R11972-5A	N/R	.035
0-80535	4150	750	37-1539	37-933	6-519-2	122-132	122-132	34R12003A	34R12003A	125-55	.045
0-80535-1	4150	750	37-1539	37-933	6-519-2						
0-80540	4150	600	37-1539	37-933	6-518-2	122-70	122-70	34R11997A	34R11997A	125-65	.028
0-80541	4150	650	37-1539	37-933	6-518-2	122-70	122-70	34R11997A	34R11997A	125-65	.028
0-80542	4150	650	37-1539	37-933	6-519-2	122-90	122-90	34R12000A	34R12000A	125-65	.055
0-80551	4160	600	703-1	N/A	6-511	122-63	34R9716-60	34R11955A	34R9716-60	125-25	.037
0-80552	4175	650	703-34	N/A	6-511	122-61	34R9716-26	34R11257A	34R9716-26	125-50	.040
0-80555	4175	650	37-1537	37-933	6-510	122-62	34R9716-54	34R11941AP	34R9716-54	125-65	.040
0-80556	4500	1150	37-1539	37-933	6-518-2	122-90	122-90	34R11972-7A	34R11972-7A	125-55	.035
0-80572	4150	700		37-933							
0-80573	4150	750		37-933							
0-80574	4150	800									
0-80575	4150	600		37-933							
0-80576	4150	750		37-933							
0-80577	4150	850		37-933							
0-80578	4500	1150		37-933							
0-80776	4150	600	37-485	37-933	6-504	122-66	122-73	34R8519AS	34R6502-3AM	125-65	.028
0-80777	4150	650	37-485	37-933	6-504	122-67	122-73	134-150	34R6497AS	125-65	.028
0-80778	4150	700	37-485	37-933	6-504	122-69	122-78	34R11174AQ	34R11176A	125-65	.028
0-80779	4150	750	37-485	37-933	6-504	122-70	122-80	34R11179AQ	34R11041AQ	125-65	.028
0-80780	4150	800	37-485	37-933	6-504	122-71	122-85	34R11196AQ	34R11198AQ	125-65	.031
0-80781	4150	850	37-485	37-933	6-504	122-80	122-78	34R11799AQ	34R9109AS	125-65 (15)	.031
0-81850	4160	600	37-119	37-933	6-506	122-66	134-9	134-128	134-9	125-65	.025
0-82010	2010	350	37-1541	N/A	6-504	122-58	N/A	N/R	N/R	125-65	.035
0-82011	2010	500	37-1541	N/A	6-504	122-80	N/A	N/R	N/R	125-65	.035
0-82012	2010	560	37-1541	N/A	6-504	122-80	N/A	N/R	N/R	125-65	.035
0-83310	4160	750	37-754	37-933	6-504	122-72	134-21	134-131	134-21	125-65	.025
0-83310-1	4160	750	37-754	37-933	6-504	122-72	134-21	134-131	134-21	125-65	.025
0-83311	4160	750	37-754	37-933	6-504	122-72	134-21	134-131	134-21	125-65	.025
0-83312	4160	750	37-754	37-933	6-504	122-72	134-21	134-131	134-21	125-65	.025
0-84010	4010	600	37-1541	N/A	6-504	122-67	122-75	N/R	N/R	125-65	.026
0-84010-1	4010	600	37-1541	N/A	6-504	122-67	122-75	N/R	N/R	125-65	.035

© **HOLLEY CARBURETORS**
VISIT <WWW.HOLLEY.COM> FOR CURRENT VERSIONS

Secondary Nozzle Size or Spring Color	Primary Bowl Gasket†	Primary Metering Block Gasket†	Secondary Bowl Gasket†	Secondary Metering Block Gasket†	Secondary Metering Plate Gasket†	Primary Fuel Bowl	Secondary Fuel Bowl	Throttle Body & Shaft Assembly	Venturi Diameter Primary	Venturi Diameter Secondary	Throttle Bore Diameter Primary	Throttle Bore Diameter Secondary
.055	108-93	108-94	108-93	108-94	N/R	34R11859	34R11857	12R11203A	1-3/8	1-3/8	1-3/4	1-3/4
.025	108-93	108-94	108-93	108-94	N/R	134-103	134-104	12R11212AS	1-1/16	1-1/16	1-7/16	1-7/16
Plain	108-83-2	108-89-2	108-90-2	108-90-2	108-27-2	134-103	134-102	N/S	1-3/8	1-7/16	1-11/16	1-11/16
Plain	108-83-2	108-89-2	108-90-2	108-90-2	108-27-2	134-103S	134-102S	12R11209AP	1-3/8	1-7/16	1-11/16	1-11/16
.029	108-93	108-94	108-93	108-94	N/R	134-103	134-104	N/S	1-9/16	1-9/16	1-11/16	1-11/16
.029	108-93	108-94	108-93	108-94	N/R	134-108	134-112	N/S	1-9/16	1-9/16	1-11/16	1-11/16
Brown	108-93	108-94	108-93	108-94	N/R	134-108	34R11442	12R11199AS	1-9/16	1-9/16	1-3/4	1-3/4
.036	108-93	108-94	108-93	108-94	N/R	134-108	134-112	12R11227A	1-9/16	1-9/16	1-3/4	1-3/4
.036	108-93	108-94	108-93	108-94	N/R	134-108	134-112	12R11227A	1-9/16	1-9/16	1-3/4	1-3/4
Brown	108-93	108-94	108-93	108-94	N/R	134-108	34R11442	12R11229A	1-9/16	1-9/16	1-3/4	1-3/4
.031	108-93	108-94	108-93	108-94	N/R	134-108	134-112	12R11234AS	1-3/8	1-3/8	1-11/16	1-11/16
.031	108-93	108-94	108-93	108-94	N/R	134-108	134-112	12R11234AS	1-3/8	1-3/8	1-11/16	1-11/16
Brown	108-93	108-94	108-93	108-94	N/R	134-108	34R12040	112-107	1-3/8	1-3/8	1-11/16	1-11/16
108-93	108-94	108-93	108-94	N/R	1-3/8	1-3/8	1-11/16	1-11/16				
Pink	108-83-2	108-89-2	108-90-2	108-89-2	N/R	34R8493A	34R11945A	12R11051AQ	1-3/8	1-3/8	1-11/16	1-11/16
.037	108-95	108-96	108-95	108-96	N/R	134-108	134-112	N/S	1.880"	1.880"	2-1/8	2-1/8
.037	108-95	108-96	108-95	108-96	N/R	134-108	134-112	N/S	1.880"	1.880"	2-1/8	2-1/8
.045	108-93	108-94	108-93	108-94	N/R	34R12043	34R12042	12R11299A	1-3/8	1-3/8	1-11/16	1-11/16
108-93	108-94	108-93	108-94	N/R	1-3/8	1-3/8	1-11/16	1-11/16				
.029	108-93	108-94	108-93	108-94	N/R	34R11843A	34R11842A	12R11290-1A	1-1/4	1-1/4	1-9/16	1-9/16
.029	108-93	108-94	108-93	108-94	N/R	34R11843A	34R11842A	12R11290-2A	1-1/4	1-1/4	1-11/16	1-11/16
.055	108-93	108-94	108-93	108-94	N/R	34R11858A	34R11847A	12R11290-3A	1-1/4	1-1/4	1-11/16	1-11/16
Red	N/A	N/A	N/A	N/A	N/A	34R11327A	34R10930A	12R11256A	1-3/8	1-3/8	1-9/16	1-9/16
Red	N/A	N/A	N/A	N/A	N/A	34R11265A	34R10930A	12R11309A	1-13/64	1-13/32	1-3/8	2
Yellow	108-92-2	108-91-2	108-90-2	108-90-2	108-27-2	34R6662-3AMP	34R7960-3AMP	12R11307A	1-13/64	1-13/32	1-3/8	2
.035	108-95	108-96	108-95	108-96	N/R	34R6063-1A	34R11417-1A	N/S	2	2	1.830	1.830
.031	108-83-2	108-89-2	108-83-2	108-89-2	N/R	134-103	134-104	12R11086A	1-1/4	1-5/16	1-9/16	1-9/16
.028	108-83-2	108-89-2	108-83-2	108-89-2	N/R	134-103	134-104	112-17	1-1/4	1-5/16	1-11/16	1-11/16
.031	108-83-2	108-89-2	108-83-2	108-89-2	N/R	134-103	134-104	12R11092A	1-5/16	1-3/8	1-11/16	1-11/16
.031	108-83-2	108-89-2	108-83-2	108-89-2	N/R	134-103	134-104	12R11147A	1-3/8	1-3/8	1-11/16	1-11/16
.031	108-83-2	108-89-2	108-83-2	108-89-2	N/R	134-103	134-104	12R11090A	1-3/8	1-7/16	1-11/16	1-11/16
.031	108-83-2	108-89-2	108-83-2	108-89-2	N/R	134-103	134-104	12R11153A	1-9/16	1-9/16	1-3/4	1-3/4
Plain	108-83-2	108-89-2	108-90-2	108-90-2	N/R	134-101	134-105	112-20	1-1/4	1-5/16	1-9/16	1-9/16
N/R	N/S	N/S	N/S	N/S	N/R	N/R	N/R	N/R	1-3/16	N/R	1-11/16	N/R
N/R	N/S	N/S	N/S	N/S	N/R	N/R	N/R	N/R	1-9/16	N/R	1-11/16	N/R
N/R	N/S	N/S	N/S	N/S	N/R	N/R	N/R	N/R	1-9/16	N/R	1-3/4	N/R
Plain	108-83-2	108-89-2	108-90-2	108-90-2	108-27-2	134-103	134-102	N/S	1-3/8	1-7/16	1-11/16	1-11/16
Plain	108-83-2	108-89-2	108-90-2	108-90-2	108-27-2	134-103	134-102	N/S	1-3/8	1-7/16	1-11/16	1-11/16
Plain	108-83-2	108-89-2	108-83-2	108-89-2	N/R	134-103	134-102	N/S	1-3/8	1-7/16	1-11/16	1-11/16
Plain	108-83-2	108-89-2	108-83-2	108-89-2	N/R	134-103	134-102	N/S	1-3/8	1-7/16	1-11/16	1-11/16
Purple	(3)	(3)	(3)	(3)	(3)	N/R	N/R	N/R	1-1/4	1-1/4	1-11/16	1-11/16
Purple	(3)	(3)	(3)	(3)	(3)	N/R	N/R	N/R	1-1/4	1-1/4	1-11/16	1-11/16

APPENDIX - HOLLEY CARBURETOR NUMERICAL LISTINGS

Carburetor Part No.	Carb. Model No.	CFM	Renew Kit	Trick Kit	Primary & Secondary Needle & Seat	Primary Main Jet	Secondary Main Jet or Plate	Primary Metering Block	Secondary Metering Block	Primary Power Valve	Primary Discharge Nozzle Size
0-84010-2	4010	600	37-1541	N/A	6-504	122-67	122-75	N/R	N/R	125-65	.035
0-84010-3	4010	600	37-1541	N/A	6-504	122-63	122-75	N/R	N/R	125-65	.035
0-84011	4010	750	37-1541	N/A	6-504	122-75	122-75	N/R	N/R	125-65 (15)	.026
0-84011-1	4010	750	37-1541	N/A	6-504	122-75	122-75	N/R	N/R	125-65 (15)	.035
0-84011-2	4010	750	37-1541	N/A	6-504	122-75	122-75	N/R	N/R	125-65	.035
0-84011-3	4010	750	37-1541	N/A	6-504	122-73	122-75	N/R	N/R	125-65	.031
0-84012	4010	600	37-1541	N/A	6-504	122-67	122-77	N/R	N/R	125-65	.026
0-84012-1	4010	600	37-1541	N/A	6-504	122-67	122-77	N/R	N/R	125-65	.035
0-84012-2	4010	600	37-1541	N/A	6-504	122-67	122-77	N/R	N/R	125-65	.035
0-84012-3	4010	600	37-1541	N/A	6-504	122-67	122-77	N/R	N/R	125-65	.031
0-84013	4010	750	37-1541	N/A	6-504	122-79	122-79	N/R	N/R	125-65 (15)	.026
0-84013-1	4010	750	37-1541	N/A	6-504	122-79	122-79	N/R	N/R	125-65 (15)	.035
0-84013-2	4010	750	37-1541	N/A	6-504	122-79	122-79	N/R	N/R	125-65 (15)	.035
0-84013-3	4010	750	37-1541	N/A	6-504	122-75	122-79	N/R	N/R	125-65	.031
0-84014	4011	650	37-1541	N/A	6-504	122-60	122-66	N/R	N/R	125-65 (15)	.026
0-84014-1	4011	650	37-1541	N/A	6-504	122-60	122-66	N/R	N/R	125-65 (15)	.026
0-84014-2	4011	650	37-1541	N/A	6-504	122-60	122-64	N/R	N/R	125-65 (15)	.026
0-84014-3	4011	650	37-1541	N/A	6-504	122-60	122-64	N/R	N/R	125-65 (15)	.026
0-84015	4011	800	37-1541	N/A	6-504	122-64	122-90	N/R	N/R	125-65 (15)	.026
0-84015-1	4011	800	37-1541	N/A	6-504	122-64	122-90	N/R	N/R	125-65 (15)	.026
0-84015-2	4011	800	37-1541	N/A	6-504	122-60	122-90	N/R	N/R	125-65 (15)	.026
0-84015-3	4011	800	37-1541	N/A	6-504	122-60	122-90	N/R	N/R	125-65 (15)	.026
0-84016	4011	650	37-1541	N/A	6-504	122-64	122-64	N/R	N/R	125-65 (15)	.026
0-84016-1	4011	650	37-1541	N/A	6-504	122-64	122-64	N/R	N/R	125-65 (15)	.026
0-84016-2	4011	650	37-1541	N/A	6-504	122-60	122-64	N/R	N/R	125-65 (15)	.026
0-84016-3	4011	650	37-1541	N/A	6-504	122-60	122-64	N/R	N/R	125-65 (15)	.026
0-84017	4011	800	37-1541	N/A	6-504	122-64	122-90	N/R	N/R	125-65 (15)	.026
0-84017-1	4011	800	37-1541	N/A	6-504	122-64	122-90	N/R	N/R	125-65 (15)	.026
0-84017-2	4011	800	37-1541	N/A	6-504	122-60	122-90	N/R	N/R	125-65 (15)	.026
0-84017-3	4011	800	37-1541	N/A	6-504	122-60	122-90	N/R	N/R	125-65 (15)	.026
0-84020	4010	600	37-1541	N/A	6-504	122-67	122-75	N/R	N/R	125-65	.026
0-84020-1	4010	600	37-1541	N/A	6-504	122-67	122-75	N/R	N/R	125-65	.035
0-84020-2	4010	600	37-1541	N/A	6-504	122-67	122-75	N/R	N/R	125-65	.035
0-84020-3	4010	600	37-1541	N/A	6-504	122-63	122-75	N/R	N/R	125-65	.035
0-84021	4011	650	37-1541	N/A	6-504	122-60	122-64	N/R	N/R	125-65 (15)	.026
0-84021-1	4011	650	37-1541	N/A	6-504	122-60	122-64	N/R	N/R	125-65 (15)	.026
0-84021-2	4011	650	37-1541	N/A	6-504	122-60	122-64	N/R	N/R	125-65 (15)	.026
0-84021-3	4011	650	37-1541	N/A	6-504	122-60	122-64	N/R	N/R	125-65 (15)	.026
0-84035	4010	600	37-1541	N/A	6-504	122-67	122-75	N/R	N/R	125-65	.035
0-84035-1	4010	600	37-1541	N/A	6-504	122-67	122-75	N/R	N/R	125-65	.035
0-84035-2	4010	600	37-1541	N/A	6-504	122-63	122-75	N/R	N/R	125-65	.035
0-84047	4010	750	37-1541	N/A	6-504	122-75	122-75	N/R	N/R	125-65 (15)	.035
0-84047-1	4010	750	37-1541	N/A	6-504	122-73	122-75	N/R	N/R	125-65 (15)	.031
0-84412	2300	500	37-474	37-933	6-504	122-73	N/R	134-137	N/R	125-50	.028
0-84776	4150	600	37-485	37-933	6-504	122-66	122-73	34R8519AS	34R6502-3AM	125-65	.028
0-84777	4150	650	37-485	37-933	6-504	122-67	122-73	134-150	34R6497AS	125-65	.028
0-84778	4150	700	37-485	37-933	6-504	122-69	122-78	N/S	N/S	125-65	.028
0-84779	4150	750	37-485	37-933	6-504	122-70	122-73	34R11179AQ	N/S	125-65	.028
0-84780	4150	800	37-485	37-933	6-504	122-71	122-85	34R11196A	N/S	125-65	.031

© **HOLLEY CARBURETORS**
VISIT <WWW.HOLLEY.COM> FOR CURRENT VERSIONS

Secondary Nozzle Size or Spring Color	Primary Bowl Gasket†	Primary Metering Block Gasket†	Secondary Bowl Gasket†	Secondary Metering Block Gasket†	Secondary Metering Plate Gasket†	Primary Fuel Bowl	Secondary Fuel Bowl	Throttle Body & Shaft Assembly	Venturi Diameter Primary	Venturi Diameter Secondary	Throttle Bore Diameter Primary	Throttle Bore Diameter Secondary
Purple	(3)	(3)	(3)	(3)	(3)	N/R	N/R	N/R	1-1/4	1-1/4	1-11/16	1-11/16
Purple	(3)	(3)	(3)	(3)	(3)	N/R	N/R	N/R	1-1/2	1-1/2	1-11/16	1-11/16
Purple	(3)	(3)	(3)	(3)	(3)	N/R	N/R	N/R	1-1/2	1-1/2	1-11/16	1-11/16
Purple	(3)	(3)	(3)	(3)	(3)	N/R	N/R	N/R	1-1/2	1-1/2	1-11/16	1-11/16
Purple	(3)	(3)	(3)	(3)	(3)	N/R	N/R	N/R	1-1/2	1-1/2	1-11/16	1-11/16
Purple	(3)	(3)	(3)	(3)	(3)	N/R	N/R	N/R	1-1/2	1-1/2	1-11/16	1-11/16
.026	(3)	(3)	(3)	(3)	(3)	N/R	N/R	N/R	1-1/4	1-1/4	1-11/16	1-11/16
.026	(3)	(3)	(3)	(3)	(3)	N/R	N/R	N/R	1-1/4	1-1/4	1-11/16	1-11/16
.026	(3)	(3)	(3)	(3)	(3)	N/R	N/R	N/R	1-1/4	1-1/4	1-11/16	1-11/16
.026	(3)	(3)	(3)	(3)	(3)	N/R	N/R	N/R	1-1/4	1-1/4	1-11/16	1-11/16
.026	(3)	(3)	(3)	(3)	(3)	N/R	N/R	N/R	1-1/2	1-1/2	1-11/16	1-11/16
.026	(3)	(3)	(3)	(3)	(3)	N/R	N/R	N/R	1-1/2	1-1/2	1-11/16	1-11/16
.026	(3)	(3)	(3)	(3)	(3)	N/R	N/R	N/R	1-1/2	1-1/2	1-11/16	1-11/16
Plain	(4)	(4)	(4)	(4)	(4)	N/R	N/R	N/R	1-5/32	1-3/8	1-3/8	2
Plain	(4)	(4)	(4)	(4)	(4)	N/R	N/R	N/R	1-5/32	1-3/8	1-3/8	2
Plain	(4)	(4)	(4)	(4)	(4)	N/R	N/R	N/R	1-5/32	1-3/8	1-3/8	2
Plain	(4)	(4)	(4)	(4)	(4)	N/R	N/R	N/R	1-5/32	1-3/8	1-3/8	2
Yellow	(4)	(4)	(4)	(4)	(4)	N/R	N/R	N/R	1-5/32	1-23/32	1-3/8	2
Yellow	(4)	(4)	(4)	(4)	(4)	N/R	N/R	N/R	1-5/32	1-23/32	1-3/8	2
Yellow	(4)	(4)	(4)	(4)	(4)	N/R	N/R	N/R	1-5/32	1-23/32	1-3/8	2
Plain	(4)	(4)	(4)	(4)	(4)	N/R	N/R	N/R	1-5/32	1-23/32	1-3/8	2
.026	(4)	(4)	(4)	(4)	(4)	N/R	N/R	N/R	1-5/32	1-3/8	1-3/8	2
.026	(4)	(4)	(4)	(4)	(4)	N/R	N/R	N/R	1-5/32	1-3/8	1-3/8	2
.026	(4)	(4)	(4)	(4)	(4)	N/R	N/R	N/R	1-5/32	1-3/8	1-3/8	2
.026	(4)	(4)	(4)	(4)	(4)	N/R	N/R	N/R	1-5/32	1-3/8	1-3/8	2
.026	(4)	(4)	(4)	(4)	(4)	N/R	N/R	N/R	1-5/32	1-23/32	1-3/8	2
.026	(4)	(4)	(4)	(4)	(4)	N/R	N/R	N/R	1-5/32	1-23/32	1-3/8	2
.026	(4)	(4)	(4)	(4)	(4)	N/R	N/R	N/R	1-5/32	1-23/32	1-3/8	2
Purple	(3)	(3)	(3)	(3)	(3)	N/R	N/R	N/R	1-1/4	1-1/4	1-11/16	1-11/16
Purple	(3)	(3)	(3)	(3)	(3)	N/R	N/R	N/R	1-1/4	1-1/4	1-11/16	1-11/16
Purple	(3)	(3)	(3)	(3)	(3)	N/R	N/R	N/R	1-1/4	1-1/4	1-11/16	1-11/16
Plain	(4)	(4)	(4)	(4)	(4)	N/R	N/R	N/R	1-5/32	1-3/8	1-3/8	2
Plain	(4)	(4)	(4)	(4)	(4)	N/R	N/R	N/R	1-5/32	1-3/8	1-3/8	2
Plain	(4)	(4)	(4)	(4)	(4)	N/R	N/R	N/R	1-5/32	1-3/8	1-3/8	2
Plain	(4)	(4)	(4)	(4)	(4)	N/R	N/R	N/R	1-5/32	1-3/8	1-3/8	2
Purple	(3)	(3)	(3)	(3)	(3)	N/R	N/R	N/R	1-1/4	1-1/4	1-11/16	1-11/16
Purple	(3)	(3)	(3)	(3)	(3)	N/R	N/R	N/R	1-1/4	1-1/4	1-11/16	1-11/16
Black	(3)	(3)	(3)	(3)	(3)	N/R	N/R	N/R	1-1/2	1-1/2	1-11/16	1-11/16
Purple	(3)	(3)	(3)	(3)	(3)	N/R	N/R	N/R	1-1/2	1-1/2	1-11/16	1-11/16
N/R	108-83-2	108-89-2	N/R	N/R	N/R	134-103	N/R	112-2	1-3/8	N/R	1-11/16	N/R
.032	108-83-2	108-89-2	108-83-2	108-89-2	N/R	134-103	134-104	12R11086A	1-1/4	1-5/16	1-9/16	1-9/16
.028	108-83-2	108-89-2	108-83-2	108-89-2	N/R	134-103	134-104	112-17	1-1/4	1-5/16	1-11/16	1-11/16
.031	108-83-2	108-89-2	108-83-2	108-89-2	N/R	134-103	134-104	12R11092A	1-5/16	1-3/8	1-11/16	1-11/16
.031	108-83-2	108-89-2	108-83-2	108-89-2	N/R	134-103	134-104	N/S	1-3/8	1-3/8	1-11/16	1-11/16
.031	108-83-2	108-89-2	108-83-2	108-89-2	N/R	134-103	134-104	12R11090A	1-3/8	1-7/16	1-11/16	1-11/16

APPENDIX - HOLLEY CARBURETOR NUMERICAL LISTINGS

Carburetor Part No.	Carb. Model No.	CFM	Renew Kit	Trick Kit	Primary & Secondary Needle & Seat	Primary Main Jet	Secondary Main Jet or Plate	Primary Metering Block	Secondary Metering Block	Primary Power Valve	Primary Discharge Nozzle Size
0-84781	4150	850	37-485	37-933	6-504	122-80	122-78	N/S	N/S	125-65	.031
0-87448	2300	350	37-1536	37-933	6-504	122-61	N/A	134-203	N/R	125-85	.031
0-89834	4160	600	37-720	37-933	6-506	122-68	134-39	N/S	134-39	125-65	.031

Footnotes
(1) 122-80 Choke Side; 122-90 Throttle Side
(2) Model 2010 Airhorn Gasket is Available Under Part Number 108-75
(3) Model 4010 Airhorn Gasket is Available Under Part Number 108-63
(4) Model 4011 Airhorn Gasket is Available Under Part Number 108-64
(5) Main Body Gasket
(12) 125-85 Secondary
(13) 125-105 Primary
(14) 125-85 Primary
(15) 125-65 Secondary
(16) 6-511 Primary
(17) 6-510 Secondary
(21) 125-65 Primary
(22) 125-35 Secondary
(24) 25R-475A-13 Early versions must use 108-29 to seal pump passage.
(29) 122-75 Diaphragm side; 122-80 Throttle Lever side
(30) 125-25 Secondary
N/A Not Available
N/S Not Serviced
N/R Not Required
†NOTE: Gasket Part Numbers now have a (-2) suffix to denote 2 gaskets per package. For example: 108-38-2.

© **HOLLEY CARBURETORS**
VISIT <WWW.HOLLEY.COM> FOR CURRENT VERSIONS

Secondary Nozzle Size or Spring Color	Primary Bowl Gasket†	Primary Metering Block Gasket†	Secondary Bowl Gasket†	Secondary Metering Block Gasket†	Secondary Metering Plate Gasket†	Primary Fuel Bowl	Secondary Fuel Bowl	Throttle Body & Shaft Assembly	Venturi Diameter Primary	Venturi Diameter Secondary	Throttle Bore Diameter Primary	Throttle Bore Diameter Secondary
.031	108-83-2	108-89-2	108-83-2	108-89-2	N/R	134-103	134-104	N/S	1-9/16	1-9/16	1-3/4	1-3/4
N/R	108-83-2	108-89-2	N/R	N/R	N/R	134-103	N/R	12R11070A	1-3/16	N/R	1-1/2	N/R
Black	108-83-2	108-91-2	108-90-2	108-90-2	108-27-2	34R8242AQ	134-105	N/S	1-1/4	1-5/16	1-9/16	1-9/16

TO ADVERTISE IN THIS SPACE CALL
0044 1305 260068
OR EMAIL INFO@VELOCE.CO.UK

TO ADVERTISE IN THIS SPACE CALL
0044 1305 260068
OR EMAIL INFO@VELOCE.CO.UK

TO ADVERTISE IN THIS SPACE CALL
0044 1305 260068
OR EMAIL INFO@VELOCE.CO.UK

Index

Accelerator pumps 27, 30, 31, 34, 37, 83, 84, 87, 88
Accelerator pump (secondary) 87, 88
Accelerator pump cams 38, 88
Accelerator pump cam positions 38
Accelerator pump check valves 40, 41
Accelerator pump diaphragms 37-39
Accelerator pump diaphragms 30cc 37-39
Accelerator pump diaphragms 50cc 37-39
Accelerator pump discharge nozzles 39, 40
Air filters 81
Automatic chokes 29

BOM sheets (what are they) 16, 17
Brass floats 34
Butterfly/throttle bore diameters 23

Carburetor sizing 24-26
Centre pivot fuel bowls 30, 33, 34
Cubic feet per minute (CFM) 7-9
CFM calculations 24, 25
CFM sizing chart 25

Chokes 29

Discharge nozzles 39, 40, 80
Discharge nozzle sizes 39, 40
Double pumper carburetors 32, 45, 46, 80, 83-88
Duracon floats 34
Dual plane inlet manifolds 89

Engine efficiency ratings 25, 26
Engine testing 81-84

Fuel bowls 30
Fuel bowl gaskets 28, 51, 52
Fuel bowl identification 30-32
Fuel bowl non-return valves 39, 40
Fuel feed transfer tube 31
Finding the List Number of a carburetor 29
Fitting metering plates 50, 51
Floats 34
Ford 7, 33, 77
Four corner idling carburetors 86

Gaskets 28, 51-52

Genuine Holley gaskets 27, 28
Getting in touch with Holley by e-mail 15, 17

Holley's technical help line 15, 17
Holley Technical Services 16, 17
Holley Performance Parts Catalog 15, 19, 28, 29, 42, 47, 48, 50, 54, 81, 84

Identification of carburetors 16-19
Idle circuitry 46, 49, 50,
Inlet manifolds 75-79, 89

Long floats 34
List Numbers 15, 17-20, 29, 30, 32, 33, 39, 43, 45, 46

Main jets 41, 42, 84, 87, 88
Main jet size chart 42
Manual chokes 29
Mechanical secondary carburetors 83-88
Metering blocks (identification of) 18-20, 34

127

Metering blocks 44-47, 84, 85
Metering block baffles 47
Metering plates 47-50, 84, 85
Metering plate (identification of) 47-49
Metering plate chart 49
Metering block gaskets 28

Needles & seats 30, 42
Needles & Seats (identification of) 43
No List Number (how to identify the carburetor) 19
Numerical Listing 15, 16, 19, 28, 80, 82, 90-109
Nytrophyl floats 34

Power valves 27, 34, 35, 80, 85
Power valve channel restrictions (PCVR) 19-23, 36, 44, 45
Power valve channel restriction hole sizes 18-20, 36, 44-46
Power valve circuitry 29
Power valve gaskets 36
Power valve rating numbers 35-36
Primary metering blocks (identification of) 18-20, 44, 45

Repair kits 2

Secondary metering blocks (identification of) 19-23, 45-47
Short secondary floats 34
Side hung fuel bowls 30-33
Sight plugs 30-33
Single plane inlet manifolds 89

Technical help from Holley 15, 17
Testing (engine) 81-84
Types of power valves 35
Throttle body and shaft assemblies 23

Vacuum secondary carburetors 82-85
Vacuum secondary diaphragm springs 52, 53, 85
Vacuum secondary diaphragms 53, 54, 80
Vacuum secondary check ball 53
Venturi diameter sizes 23

Visit Veloce on the web - www.veloce.co.uk